Division

Workbook

ANSWER KEY can be found at www.preschoolprepco.com/answers

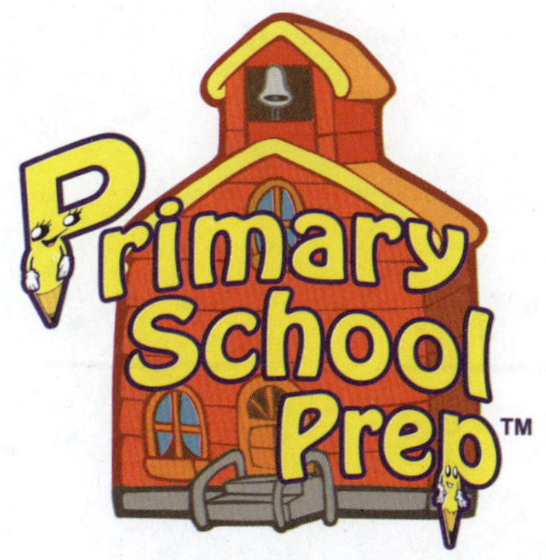

A Divison of Preschool Prep Company™

866-451-5600
www.PreschoolPrepCo.com
P.O. Box 1159, Danville CA 94526

Fill in the Answer

Fill in the blanks to complete the equations.

$1 \div \_\_ = 1$ $\quad$ $\_\_ \div 1 = 1$ $\quad$ $1 \div 1 = \_\_$

$1 \div 1 = \_\_$ $\quad$ $1 \div \_\_ = 1$ $\quad$ $1 \div \_\_ = 1$

$1 \div \_\_ = 1$ $\quad$ $\_\_ \div 1 = 1$ $\quad$ $\_\_ \div 1 = 1$

$\_\_ \div 1 = 1$ $\quad$ $1 \div 1 = \_\_$ $\quad$ $1 \div 1 = \_\_$

$\begin{array}{r} 1 \\ \div\ 1 \\ \hline \_\_ \end{array}$ $\quad$ $\begin{array}{r} 1 \\ \div\ \_\_ \\ \hline 1 \end{array}$ $\quad$ $\begin{array}{r} \_\_ \\ \div\ 1 \\ \hline 1 \end{array}$ $\quad$ $\begin{array}{r} \_\_ \\ \div\ 1 \\ \hline 1 \end{array}$ $\quad$ $\begin{array}{r} 1 \\ \div\ 1 \\ \hline \_\_ \end{array}$ $\quad$ $\begin{array}{r} \_\_ \\ \div\ 1 \\ \hline 1 \end{array}$

$1\overline{)1}$ (quotient: __) $\quad$ $\frac{1}{1} = \_\_$ $\quad$ $1\overline{)1}$ (quotient: __) $\quad$ $\frac{1}{1} = \_\_$ $\quad$ $1\overline{)1}$ (quotient: __)

Activities

There is one paint brush.

There is one child.

How many brushes can he have? ______

Use the numbers below to make equations.

1 1 1

$1 \div \square = \square$

$\square \div 1 = \square$

$1 \div \square = \square$

$\square \div \square = 1$

Make equations with the numbers below.

1 1 1

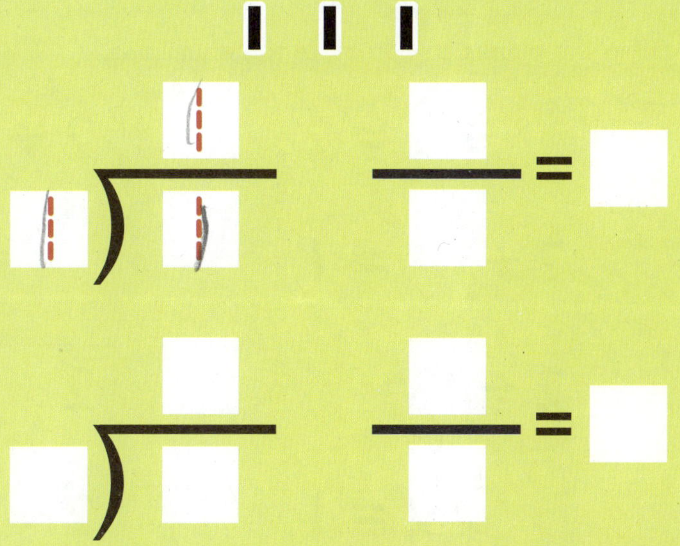

Solve the problems to get the girl to the easel.

Fill in the Answer

Fill in the blanks to complete the equations.

$\_\_\_ \div 1 = 2$　　$2 \div \_\_\_ = 2$　　$2 \div 1 = \_\_\_$

$2 \div \_\_\_ = 1$　　$\_\_\_ \div 1 = 2$　　$2 \div \_\_\_ = 2$

$2 \div 1 = \_\_\_$　　$2 \div 2 = \_\_\_$　　$\_\_\_ \div 2 = 1$

$\_\_\_ \div 2 = 1$　　$2 \div \_\_\_ = 1$　　$2 \div 2 = \_\_\_$

2	2	___	2	2	2
÷ ___	÷ 1	÷ 1	÷ 2	÷ ___	÷ ___
1	___	2	___	2	1

$1\overline{)2}$ (quotient: ___)　　$\frac{2}{2} = \_\_\_$　　$2\overline{)2}$ (quotient: ___)　　$\frac{2}{1} = \_\_\_$　　$1\overline{)2}$ (quotient: ___)

Activities

There are two toothbrushes.

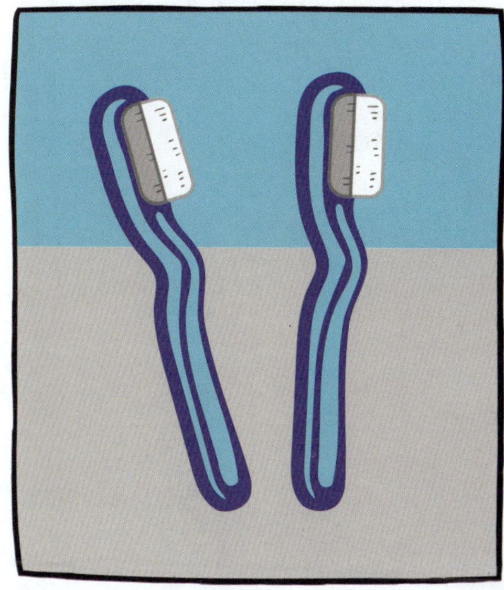

There is one cup.

How many brushes can you put in the cup? 2

Use the numbers below to make equations.

1 2 2

$2 \div 1 = 2$

$1 \div 2 = 2$

$2 \div 1 = 3$

$2 \div 1 = 2$

Make equations with the numbers below.

1 2 2

$1 \overline{)2}$ = 2

$\frac{1}{2} = 2$

$2 \overline{)2}$ = 1

$\frac{2}{2} = 1$

Solve the problems to get the dinosaur to the toothbrush.

$2 \div 1 = 2$

$2 \div 2 = 1$

$2 \div 1 = 2$

$2 \div 2 = 1$

$2 \div 2 = 1$

Fill in the Answer

Fill in the blanks to complete the equations.

3 ÷ ___ = 3	3 ÷ 3 = ___	___ ÷ 3 = 1
___ ÷ 3 = 1	3 ÷ ___ = 3	3 ÷ 3 = ___
3 ÷ 1 = ___	___ ÷ 1 = 3	___ ÷ 1 = 3
3 ÷ ___ = 1	3 ÷ 1 = ___	3 ÷ ___ = 1

$$\begin{array}{r}3\\ \div\ 1\\ \hline \_\_\end{array}\qquad \begin{array}{r}\_\_\\ \div\ 3\\ \hline 1\end{array}\qquad \begin{array}{r}\_\_\\ \div\ 1\\ \hline 3\end{array}\qquad \begin{array}{r}\_\_\\ \div\ 3\\ \hline 1\end{array}\qquad \begin{array}{r}3\\ \div\ \_\_\\ \hline 3\end{array}\qquad \begin{array}{r}3\\ \div\ 3\\ \hline \_\_\end{array}$$

$$1\overline{)3}^{\,\_\_}\qquad \frac{3}{3} = \_\_\qquad 3\overline{)3}^{\,\_\_}\qquad \frac{3}{1} = \_\_\qquad 1\overline{)3}^{\,\_\_}$$

Activities

There are three streams.

There are three logs.

When divided evenly, how many logs will be in each stream? ____

Use the numbers below to make equations.

1 3 3

$\square \div \square = 3$

$\square \div 3 = \square$

$3 \div \square = \square$

$\square \div \square = 1$

Make equations with the numbers below.

1 3 3

$\square \overline{\smash{)}\square}$ with quotient $\square$ $\dfrac{\square}{\square} = \square$

$\square \overline{\smash{)}\square}$ with quotient $\square$ $\dfrac{\square}{\square} = \square$

Solve the problems to get the beaver to the stream.

$3 \div 1 = \square$

$3 \div 3 = \square$

$\square \div 1 = 3$

$3 \div 3 = \square$

$3 \div \square = 1$

$3 \div 1 = \square$

Fill in the Answer

Fill in the blanks to complete the equations.

$4 \div 1 = 4$ $4 \div 4 = 1$ $4 \div 4 = 1$

$4 \div 1 = 4$ $4 \div 4 = 1$ $4 \div 1 = 4$

$4 \div 1 = 4$ $4 \div 1 = 4$ $4 \div 4 = 1$

$4 \div 4 = 1$ $4 \div 1 = 4$ $4 \div 4 = 1$

$$\begin{array}{r} 4 \\ \div\ 4 \\ \hline 1 \end{array} \quad \begin{array}{r} 4 \\ \div\ 4 \\ \hline 1 \end{array} \quad \begin{array}{r} 4 \\ \div\ 4 \\ \hline 1 \end{array} \quad \begin{array}{r} 4 \\ \div\ 1 \\ \hline 4 \end{array} \quad \begin{array}{r} 4 \\ \div\ 1 \\ \hline 4 \end{array} \quad \begin{array}{r} 4 \\ \div\ 1 \\ \hline 4 \end{array}$$

$$\begin{array}{r} 4 \\ 1\overline{)4} \end{array} \quad \frac{4}{4} = 1 \quad \begin{array}{r} 1 \\ 4\overline{)4} \end{array} \quad \frac{4}{1} = 4 \quad \begin{array}{r} 1 \\ 4\overline{)4} \end{array}$$

Activities

There are four coins.

There are four machines.

How many coins are there for each machine when divided evenly? ____

Use the numbers below to make equations.

1 4 4

$\square \div \square = 4$

$4 \div \square = \square$

$\square \div 1 = \square$

$\square \div 4 = \square$

Make equations with the numbers below.

1 4 4

Solve the problems to get the girl to the coin machine.

Fill in the Answer

Fill in the blanks to complete the equations.

$5 \div 5 = 1$ $5 \div 5 = 1$ $5 \div 5 = 1$

$5 \div 1 = 5$ $5 \div 1 = 5$ $5 \div 5 = 1$

$5 \div 5 = 1$ $5 \div 5 = 1$ $5 \div 1 = 5$

$5 \div 1 = 5$ $5 \div 1 = 5$ $5 \div 1 = 5$

$$\begin{array}{r} 5 \\ \div\ 1 \\ \hline 5 \end{array} \quad \begin{array}{r} 5 \\ \div\ 1 \\ \hline 5 \end{array} \quad \begin{array}{r} 5 \\ \div\ 1 \\ \hline 5 \end{array} \quad \begin{array}{r} 5 \\ \div\ 5 \\ \hline 1 \end{array} \quad \begin{array}{r} 5 \\ \div\ 5 \\ \hline 1 \end{array} \quad \begin{array}{r} 5 \\ \div\ 1 \\ \hline 5 \end{array}$$

$$1\overline{)5} = 5 \quad \frac{5}{5} = 1 \quad 5\overline{)5} = 1 \quad \frac{5}{1} = 5 \quad 1\overline{)5} = 5$$

Activities

There are five ice cream bars.

There are five children.

How many bars does each child get when divided evenly? ______

Use the numbers below to make equations.

1 5 5

___ ÷ 1 = ___

___ ÷ 5 = ___

___ ÷ ___ = 5

5 ÷ ___ = ___

Make equations with the numbers below.

1 5 5

Solve the problems to get the lizard to the ice cream stand.

Fill in the Answer

Fill in the blanks to complete the equations.

$6 \div 6 = \_\_\_$ $\qquad$ $6 \div \_\_\_ = 1$ $\qquad$ $\_\_\_ \div 6 = 1$

$\_\_\_ \div 1 = 6$ $\qquad$ $\_\_\_ \div 6 = 1$ $\qquad$ $6 \div \_\_\_ = 1$

$6 \div \_\_\_ = 6$ $\qquad$ $6 \div 6 = \_\_\_$ $\qquad$ $6 \div 1 = \_\_\_$

$6 \div 1 = \_\_\_$ $\qquad$ $6 \div \_\_\_ = 6$ $\qquad$ $\_\_\_ \div 1 = 6$

$$\begin{array}{r}6\\ \div\ \_\_\\ \hline 1\end{array} \qquad \begin{array}{r}6\\ \div\ 1\\ \hline \_\_\end{array} \qquad \begin{array}{r}\_\_\\ \div\ 6\\ \hline 1\end{array} \qquad \begin{array}{r}6\\ \div\ 6\\ \hline \_\_\end{array} \qquad \begin{array}{r}\_\_\\ \div\ 1\\ \hline 6\end{array} \qquad \begin{array}{r}6\\ \div\ \_\_\\ \hline 6\end{array}$$

$$\begin{array}{r}\_\_\\ 1\overline{)6}\end{array} \qquad \frac{6}{6} = \_\_\_ \qquad \begin{array}{r}\_\_\\ 6\overline{)6}\end{array} \qquad \frac{6}{1} = \_\_\_ \qquad \begin{array}{r}\_\_\\ 1\overline{)6}\end{array}$$

Activities

There are six fishing hooks.

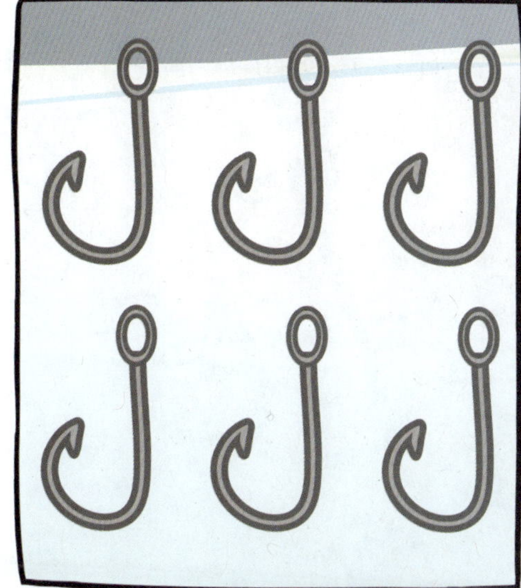

There are six fishing poles.

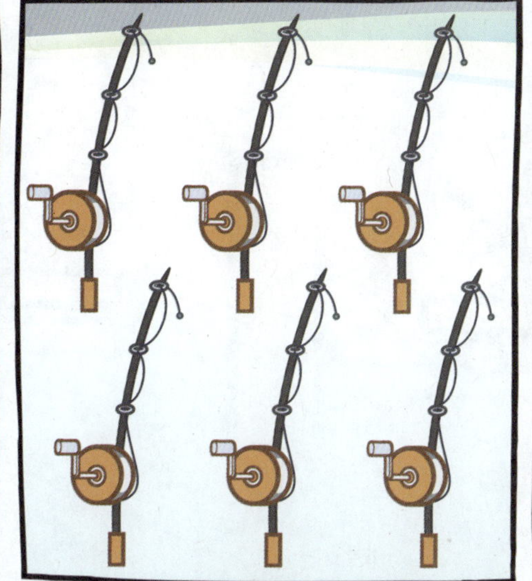

How many hooks can we put on each pole when dividing evenly? 6

Use the numbers below to make equations.

1 6 6

6 ÷ 1 = **6**

6 ÷ **6** = 1

6 ÷ 6 = 1

6 ÷ **1** = 6

Make equations with the numbers below.

1 6 6

$1\overline{)6}$ = 6 $\frac{6}{1} = 6$

$6\overline{)6}$ = 1 $\frac{6}{6} = 1$

Solve the problems to get the seal to the fishing hole.

6÷6= 1

6÷ 6 =1

6÷6= 1

6÷1= 6

6 ÷1=6

Fill in the Answer

Fill in the blanks to complete the equations.

$7 \div 7 = 1$ $\quad$ $7 \div 1 = 7$ $\quad$ $7 \div 7 = 1$

$7 \div 1 = 7$ $\quad$ $7 \div 7 = 1$ $\quad$ $7 \div 7 = 1$

$7 \div 7 = 1$ $\quad$ $7 \div 7 = 1$ $\quad$ $7 \div 1 = 7$

$7 \div 1 = 7$ $\quad$ $7 \div 1 = 7$ $\quad$ $7 \div 1 = 7$

$$\begin{array}{r} 7 \\ \div\ 7 \\ \hline 1 \end{array} \quad \begin{array}{r} 7 \\ \div\ 7 \\ \hline 1 \end{array} \quad \begin{array}{r} 7 \\ \div\ 1 \\ \hline 7 \end{array} \quad \begin{array}{r} 7 \\ \div\ 1 \\ \hline 7 \end{array} \quad \begin{array}{r} 7 \\ \div\ 1 \\ \hline 7 \end{array} \quad \begin{array}{r} 7 \\ \div\ 7 \\ \hline 1 \end{array}$$

$$1\overline{)7}\ \text{(quotient 7)} \quad \frac{7}{7} = 1 \quad 7\overline{)7}\ \text{(quotient 1)} \quad \frac{7}{1} = 7 \quad 7\overline{)7}\ \text{(quotient 1)}$$

Activities

There are seven nails.

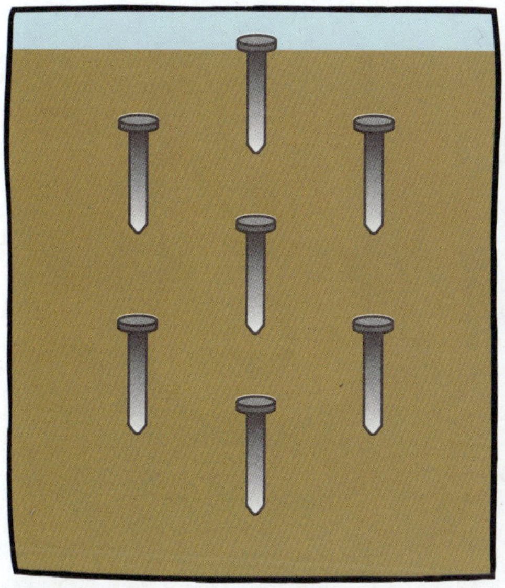

There is one box.

We can use ___7___ nails to fix the box.

Use the numbers below to make equations.

1 7 7

7 ÷ 1 = 7

7 ÷ 7 = 1

7 ÷ 7 = 1

7 ÷ 1 = 7

Make equations with the numbers below.

1 7 7

Solve the problems to get the construction worker to his hammer.

7÷7= 1

7 ÷1=7

7÷1= 7

7÷7= 1

7÷ 7 =1

Fill in the Answer

Fill in the blanks to complete the equations.

___ ÷ 1 = 8	8 ÷ 1 = ___	___ ÷ 1 = 8
8 ÷ 8 = ___	8 ÷ ___ = 8	8 ÷ 1 = ___
8 ÷ ___ = 1	___ ÷ 8 = 1	___ ÷ 8 = 1
8 ÷ ___ = 8	8 ÷ 8 = ___	8 ÷ ___ = 1

$$\begin{array}{r} 8 \\ \div\ \_\_ \\ \hline 8 \end{array} \quad \begin{array}{r} 8 \\ \div\ 8 \\ \hline \_\_ \end{array} \quad \begin{array}{r} \_\_ \\ \div\ 1 \\ \hline 8 \end{array} \quad \begin{array}{r} 8 \\ \div\ 1 \\ \hline \_\_ \end{array} \quad \begin{array}{r} 8 \\ \div\ \_\_ \\ \hline 1 \end{array} \quad \begin{array}{r} 8 \\ \div\ \_\_ \\ \hline 8 \end{array}$$

$$1\overline{)8} = \_\_ \quad \frac{8}{1} = \_\_ \quad 8\overline{)8} = \_\_ \quad \frac{8}{8} = \_\_ \quad 1\overline{)8} = \_\_$$

Activities

There are eight cookies.

There is one plate.

Using all of the cookies, how many can we put on the plate? ____

Use the numbers below to make equations.

1 8 8

$\_\_ \div \_\_ = 8$

$8 \div \_\_ = \_\_$

$\_\_ \div 8 = \_\_$

$\_\_ \div 1 = \_\_$

Make equations with the numbers below.

1 8 8

Solve the equations to get the panda to the bamboo.

Fill in the Answer

Fill in the blanks to complete the equations.

$9 \div 1 = 9$	$9 \div 9 = 1$	$9 \div 1 = 9$
$9 \div 1 = 9$	$9 \div 1 = 9$	$9 \div 9 = 1$
$9 \div 9 = 1$	$9 \div 1 = 9$	$9 \div 1 = 9$
$9 \div 9 = 1$	$9 \div 9 = 1$	$9 \div 9 = 1$

$9 \div 9 = 1$ $9 \div 9 = 1$ $9 \div 9 = 1$ $9 \div 1 = 9$ $9 \div 1 = 9$ $9 \div 1 = 9$

$1\overline{)9}$ with answer 9 $\frac{9}{9} = 9$ $9\overline{)9}$ with answer 1 $\frac{9}{1} = 9$ $1\overline{)9}$ with answer 9

Activities

There are nine fireflies.

There are nine leaves.

How many fireflies will sit on each leaf when divided evenly? ____

Use the numbers below to make equations.

Make equations with the numbers below.

Solve the problems to get the owl to his clock.

$9 \div 1 = \square$

$\square \div 9 = 1$

$9 \div 9 = \square$

$9 \div \square = 9$

Fill in the Answer

Fill in the blanks to complete the equations.

$10 \div 1 = 10$	$10 \div 10 = 1$	$10 \div 10 = 1$
$10 \div 10 = 1$	$10 \div 10 = 1$	$10 \div 10 = 1$
$10 \div 1 = 10$	$10 \div 1 = 10$	$10 \div 10 = 1$
$10 \div 1 = 10$	$10 \div 1 = 10$	$10 \div 1 = 10$

$$\begin{array}{r}10\\ \div\ 1\\ \hline 10\end{array} \quad \begin{array}{r}10\\ \div\ 1\\ \hline 10\end{array} \quad \begin{array}{r}10\\ \div\ 10\\ \hline 1\end{array} \quad \begin{array}{r}10\\ \div\ 10\\ \hline 1\end{array} \quad \begin{array}{r}10\\ \div\ 1\\ \hline 10\end{array} \quad \begin{array}{r}10\\ \div\ 10\\ \hline 1\end{array}$$

$$1\overline{)10} = 10 \qquad \frac{10}{1} = 10 \qquad 10\overline{)10} = 1 \qquad \frac{10}{10} = 1 \qquad 1\overline{)10} = 10$$

Activities

There are ten slices of pepperoni.

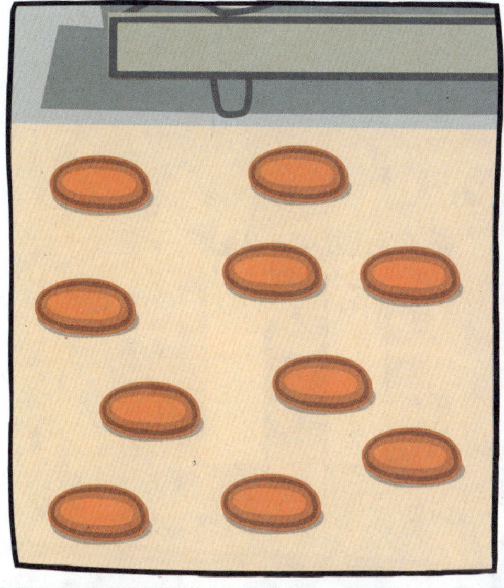

There is one bowl.

How many slices of pepperoni can we put in the bowl? ______

Use the numbers below to make equations.

1 10 10

Make equations with the numbers below.

1 10 10

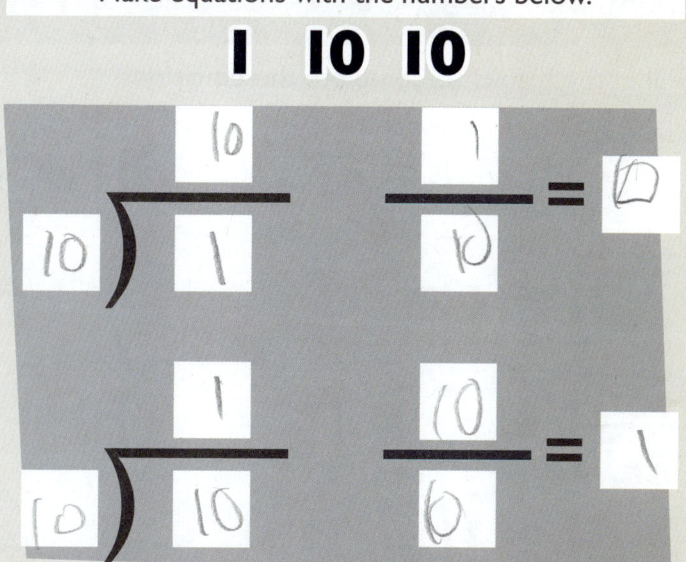

Solve the equations to get the chef to his brick oven.

$10 \div 1 = \_\_\_$

$\_\_\_ \div 10 = 1$

$10 \div \_\_\_ = 1$

$10 \div 10 = \_\_\_$

$\_\_\_$

$\div\ 1$

10

Fill in the Answer

Fill in the blanks to complete the equations.

___ ÷ I = II　　II ÷ I = ___　　II ÷ II = ___

II ÷ I = ___　　___ ÷ II = I　　___ ÷ II = I

II ÷ ___ = I　　II ÷ ___ = II　　II ÷ ___ = I

II ÷ II = ___　　___ ÷ I = II　　II ÷ ___ = II

II	___	II	II	II	___
÷ I	÷ I	÷ ___	÷ II	÷ ___	÷ II
___	II	I	___	II	I

I) II (quotient: ___)　　II / I = ___　　II) II (quotient: ___)　　II / II = ___　　II) II (quotient: ___)

Activities

There are eleven candles.

There is one cake.

How many candles can we put on the cake? ______

Use the numbers below to make equations.

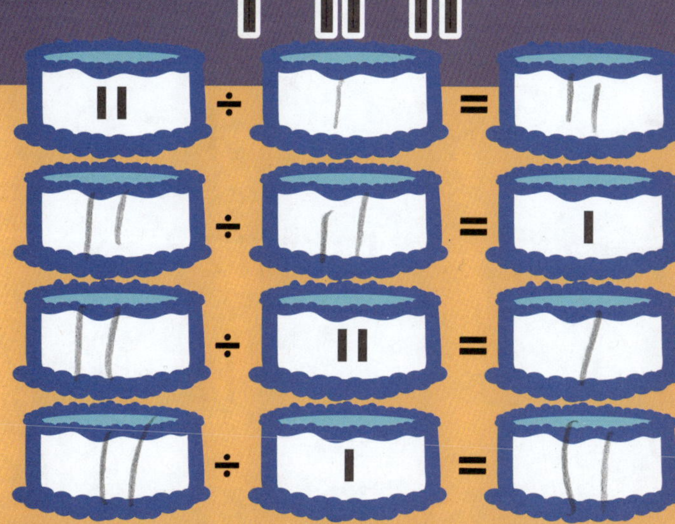

Make equations with the numbers below.

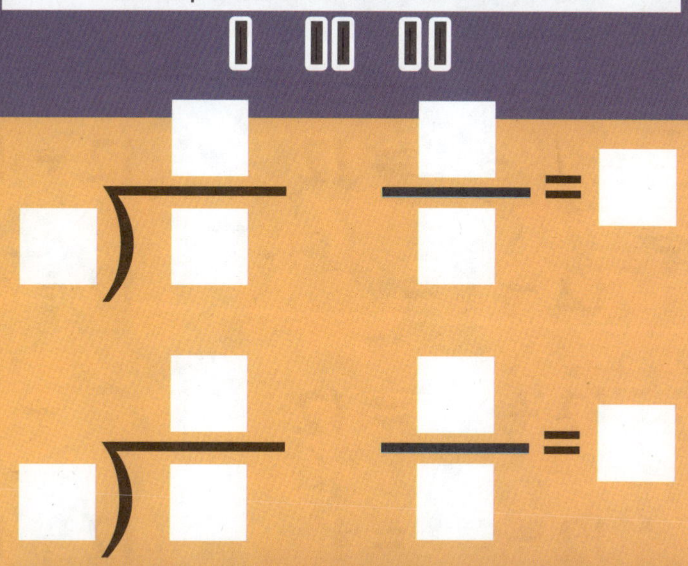

Solve the equations to get the firefighter to the fire.

II ÷ ☐ = I

II ÷ I = ☐

☐ ÷ II = I

II ÷ I = ☐

II ÷ ☐ = I

Fill in the Answer

Fill in the blanks to complete the equations.

$12 \div 1 = 12$	$12 \div 12 = 1$	$12 \div 1 = 12$
$12 \div 1 = 12$	$12 \div 1 = 12$	$12 \div 12 = 1$
$12 \div 1 = 12$	$12 \div 12 = 1$	$12 \div 12 = 1$
$12 \div 12 = 1$	$12 \div 1 = 12$	$12 \div 12 = 1$

$$\begin{array}{r}12\\ \div\ 1\\ \hline 12\end{array}\quad \begin{array}{r}12\\ \div\ 1\\ \hline 12\end{array}\quad \begin{array}{r}12\\ \div\ 1\\ \hline 12\end{array}\quad \begin{array}{r}12\\ \div\ 12\\ \hline 1\end{array}\quad \begin{array}{r}12\\ \div\ 12\\ \hline 1\end{array}\quad \begin{array}{r}12\\ \div\ 1\\ \hline 12\end{array}$$

$$1\overline{)12}^{\,12} \qquad \frac{12}{1} = 12 \qquad 12\overline{)12}^{\,1} \qquad \frac{12}{12} = 1 \qquad 1\overline{)12}^{\,12}$$

Activities

There are twelve balls.

There is one basket.

Using all of the balls, how many are there for the basket? ____

Use the numbers below to make equations.

1 12 12

$\square \div 12 = \square$

$12 \div \square = \square$

$\square \div \square = 1$

$\square \div 1 = \square$

Make equations with the numbers below.

1 12 12

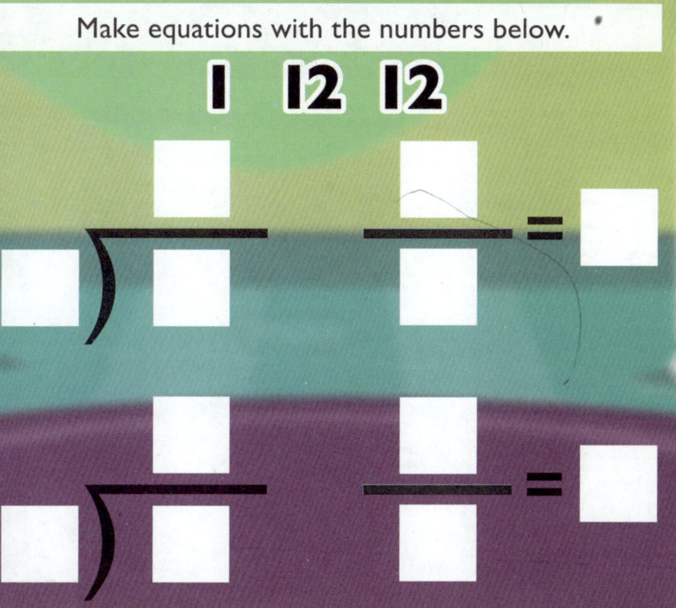

Solve the equations to get the baby to his bottle.

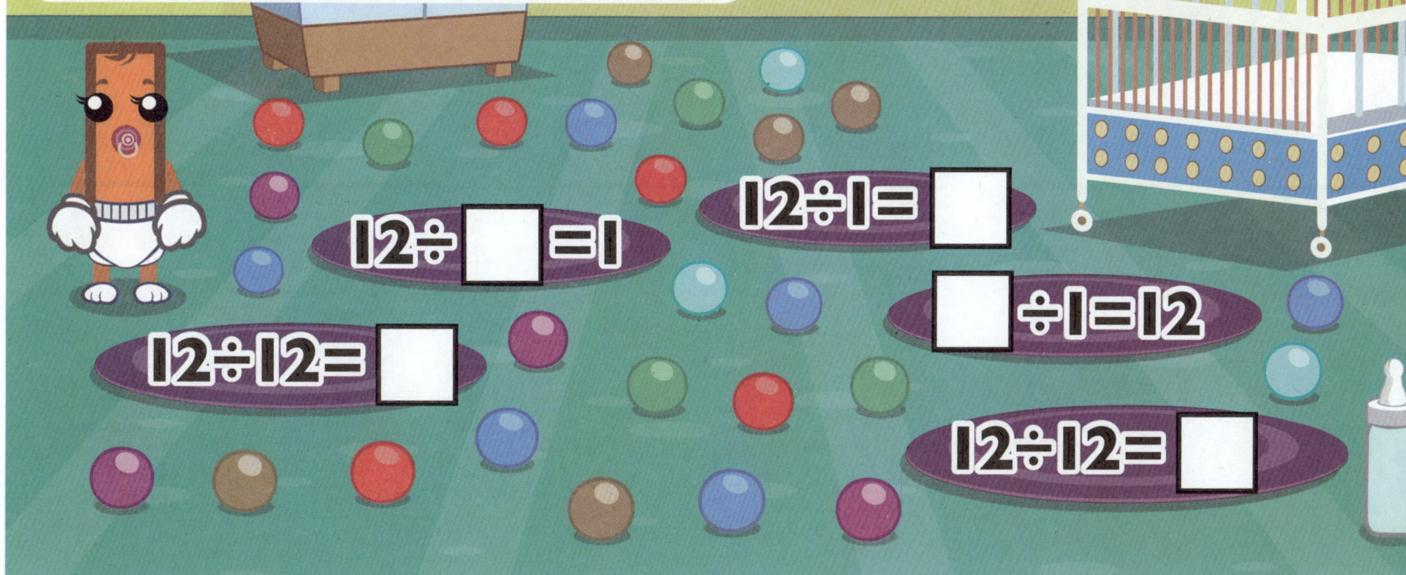

Fill in the Answer

Fill in the blanks to complete the equations.

$4 \div 2 = 2$ $4 \div 2 = 2$ $4 \div 2 = 2$

$4 \div 2 = 2$ $4 \div 2 = 2$ $4 \div 2 = 2$

$4 \div 2 = 2$ $4 \div 2 = 2$ $4 \div 2 = 2$

$4 \div 2 = 2$ $4 \div 2 = 2$ $4 \div 2 = 2$

$\begin{array}{r} 4 \\ \div\ 2 \\ \hline 2 \end{array}$ $\begin{array}{r} 4 \\ \div\ 2 \\ \hline 2 \end{array}$ $\begin{array}{r} 4 \\ \div\ 2 \\ \hline 2 \end{array}$ $\begin{array}{r} 4 \\ \div\ 2 \\ \hline 2 \end{array}$ $\begin{array}{r} 4 \\ \div\ 2 \\ \hline 2 \end{array}$ $\begin{array}{r} 4 \\ \div\ 2 \\ \hline 2 \end{array}$

$2 \overline{)4}$ with quotient 2 $\frac{4}{2} = 2$ $2 \overline{)4}$ with quotient 4 $\frac{4}{2} = 2$ $2 \overline{)4}$ with quotient 4

Activities

There are four birds.

There are two ears of corn.

How many birds will share each ear of corn? _____

Use the numbers below to make equations.

2 2 4

___ ÷ 2 = ___

___ ÷ ___ = 2

4 ÷ ___ = ___

___ ÷ 2 = ___

Make equations with the numbers below.

2 2 4

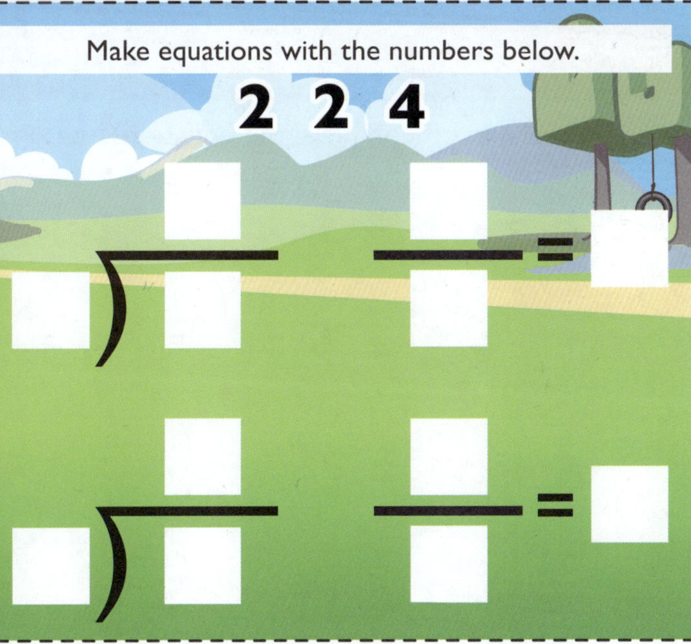

Divide the corn into two even groups by circling each group.

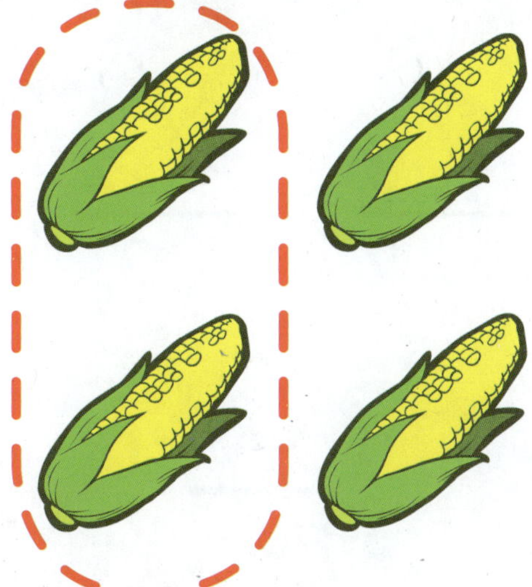

Write the equation to match the picture. ______ ÷ ______ = ______

Fill in the Answer

Fill in the blanks to complete the equations.

$6 \div 2 = 3$	$6 \div 3 = 2$	$6 \div 2 = 3$
$6 \div 2 = 3$	$6 \div 3 = 2$	$6 \div 2 = 3$
$6 \div 3 = 2$	$6 \div 2 = 3$	$6 \div 3 = 2$
$6 \div 2 = 3$	$6 \div 3 = 2$	$6 \div 2 = 3$

$$\begin{array}{r}6\\ \div\ 3\\ \hline 2\end{array} \quad \begin{array}{r}6\\ \div\ 3\\ \hline 2\end{array} \quad \begin{array}{r}6\\ \div\ 3\\ \hline 2\end{array} \quad \begin{array}{r}6\\ \div\ 2\\ \hline 3\end{array} \quad \begin{array}{r}6\\ \div\ 2\\ \hline 3\end{array} \quad \begin{array}{r}6\\ \div\ 3\\ \hline 2\end{array}$$

$$2\overline{)6}\ \text{answer: } 3 \qquad \frac{6}{3} = 2 \qquad 3\overline{)6}\ \text{answer: } 2 \qquad \frac{6}{2} = 3 \qquad 2\overline{)6}\ \text{answer: } 3$$

Activities

There are six coins.

There are three treasure boxes.

How many coins can we divide evenly into each box? ______

Use the numbers below to make equations.

2 3 6

6 ÷ ◯ = ◯

◯ ÷ ◯ = 2

◯ ÷ 2 = ◯

◯ ÷ ◯ = 3

Make equations with the numbers below.

2 3 6

Divide the fishbowls into three even groups by circling each group.

Write the equation to match the picture. ______ ÷ ______ = ______

Fill in the Answer

Fill in the blanks to complete the equations.

$8 \div 2 = 4$	$8 \div 2 = 4$	$8 \div 4 = 2$
$8 \div 4 = 2$	$8 \div 2 = 4$	$8 \div 4 = 2$
$8 \div 4 = 2$	$8 \div 2 = 4$	$8 \div 2 = 4$
$8 \div 4 = 2$	$8 \div 4 = 2$	$8 \div 2 = 4$

$$\begin{array}{r} 8 \\ \div\ \_\_\_ \\ \hline 4 \end{array} \qquad \begin{array}{r} 8 \\ \div\ 2 \\ \hline \_\_\_ \end{array} \qquad \begin{array}{r} \_\_\_ \\ \div\ 2 \\ \hline 4 \end{array} \qquad \begin{array}{r} 8 \\ \div\ 4 \\ \hline \_\_\_ \end{array} \qquad \begin{array}{r} \_\_\_ \\ \div\ 4 \\ \hline 2 \end{array} \qquad \begin{array}{r} 8 \\ \div\ \_\_\_ \\ \hline 2 \end{array}$$

$$\begin{array}{r} \_\_\_ \\ 4 \overline{)8} \end{array} \qquad \frac{8}{4} = \_\_\_ \qquad \begin{array}{r} \_\_\_ \\ 2 \overline{)8} \end{array} \qquad \frac{8}{2} = \_\_\_ \qquad \begin{array}{r} \_\_\_ \\ 4 \overline{)8} \end{array}$$

Activities

There are eight buttons.

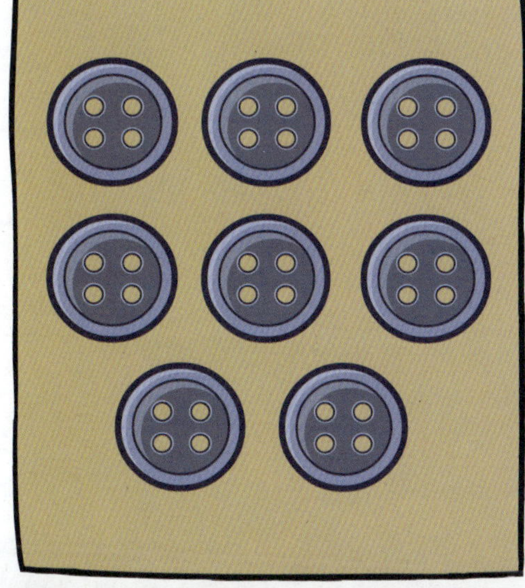

There are two shirts.

Dividing evenly, how many buttons can we sew on each shirt? _____

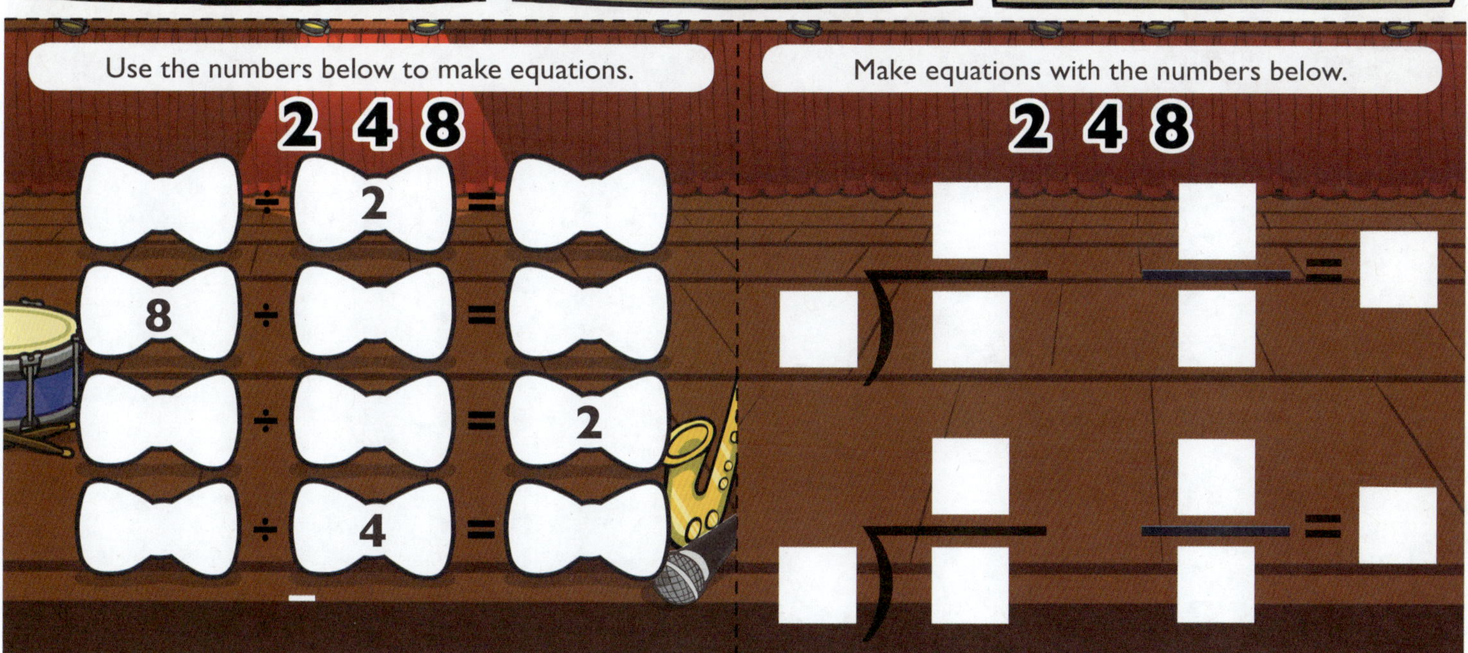

Divide the irons into four even groups by circling each group.

Write the equation to match the picture. _____ ÷ _____ = _____

Fill in the Answer

Fill in the blanks to complete the equations.

10 ÷ ___ = 2	___ ÷ 5 = 2	10 ÷ 2 = ___
10 ÷ 5 = ___	10 ÷ ___ = 2	10 ÷ ___ = 5
10 ÷ ___ = 2	___ ÷ 2 = 5	___ ÷ 5 = 2
___ ÷ 2 = 5	10 ÷ 2 = ___	10 ÷ 5 = ___

10	10	10	___	___	10
÷ 2	÷ ___	÷ 5	÷ 2	÷ 5	÷ 2
___	5	___	5	2	___

$5\overline{)10}$ (quotient: ___) $\frac{10}{5}$ = ___ $2\overline{)10}$ (quotient: ___) $\frac{10}{2}$ = ___ $5\overline{)10}$ (quotient: ___)

Activities

There are ten gems.

There are two crowns.

How many gems should we put on each crown when dividing evenly? _____

Use the numbers below to make equations.

2 5 10

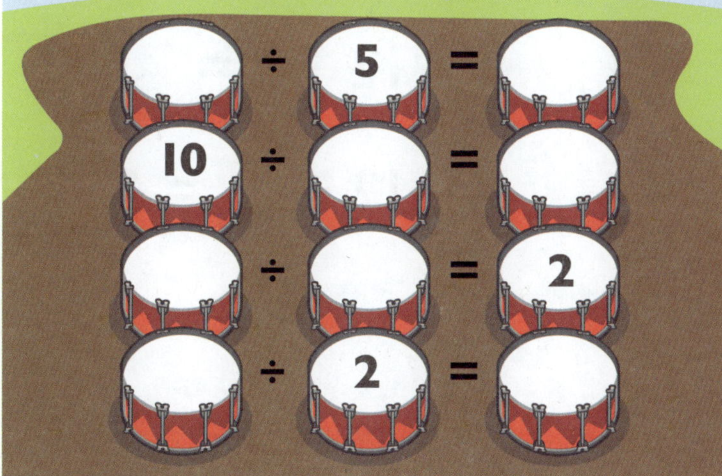

Make equations with the numbers below.

2 5 10

Divide the drums into five even groups by circling each group.

Write the equation to match the picture. ______ ÷ ______ = ______

Fill in the Answer

Fill in the blanks to complete the equations.

$\square \div 2 = 6$	$12 \div \square = 6$	$12 \div 2 = \square$
$12 \div \square = 2$	$\square \div 2 = 6$	$12 \div \square = 6$
$12 \div 2 = \square$	$12 \div 6 = \square$	$\square \div 6 = 2$
$\square \div 2 = 6$	$12 \div \square = 2$	$12 \div 6 = \square$

$$\begin{array}{r} 12 \\ \div\ \square \\ \hline 6 \end{array} \qquad \begin{array}{r} 12 \\ \div\ 6 \\ \hline \square \end{array} \qquad \begin{array}{r} \square \\ \div\ 6 \\ \hline 2 \end{array} \qquad \begin{array}{r} 12 \\ \div\ 2 \\ \hline \square \end{array} \qquad \begin{array}{r} \square \\ \div\ 2 \\ \hline 6 \end{array} \qquad \begin{array}{r} \square \\ \div\ 6 \\ \hline 2 \end{array}$$

$$\begin{array}{r} \square \\ 6\overline{)12} \end{array} \qquad \frac{12}{2} = \square \qquad \begin{array}{r} \square \\ 2\overline{)12} \end{array} \qquad \frac{12}{6} = \square \qquad \begin{array}{r} \square \\ 6\overline{)12} \end{array}$$

Activities

There are twelve arrows.

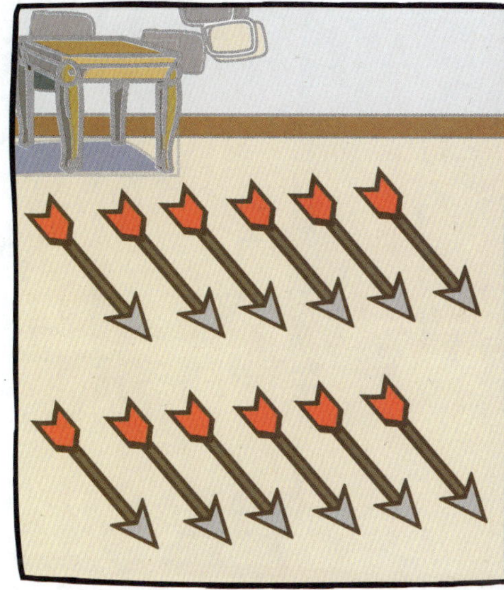

There are two quivers.

How many arrows can be divided evenly into each quiver? _____

Use the numbers below to make equations.

÷ = 6

÷ 6 =

Make equations with the numbers below.

Divide the apples into six equal groups by circling each group.

Write the equation to match the picture. ______ ÷ ______ = ______

Fill in the Answer

Fill in the blanks to complete the equations.

$14 \div \square = 7$ $14 \div 7 = \square$ $\square \div 7 = 2$

$\square \div 2 = 7$ $14 \div \square = 7$ $14 \div 7 = \square$

$14 \div 2 = \square$ $\square \div 7 = 2$ $\square \div 2 = 7$

$14 \div \square = 2$ $14 \div 2 = \square$ $14 \div \square = 2$

$\begin{array}{r} 14 \\ \div\ 7 \\ \hline \square \end{array}$ $\begin{array}{r} \square \\ \div\ 7 \\ \hline 2 \end{array}$ $\begin{array}{r} \square \\ \div\ 2 \\ \hline 7 \end{array}$ $\begin{array}{r} \square \\ \div\ 7 \\ \hline 2 \end{array}$ $\begin{array}{r} 14 \\ \div\ \square \\ \hline 2 \end{array}$ $\begin{array}{r} 14 \\ \div\ 2 \\ \hline \square \end{array}$

$7\overline{)14}$ with $\square$ above $\frac{14}{7} = \square$ $2\overline{)14}$ with $\square$ above $\frac{14}{2} = \square$ $7\overline{)14}$ with $\square$ above

Activities

There are fourteen seats.

The seats are placed into two even rows.

How many seats are there in each row? __

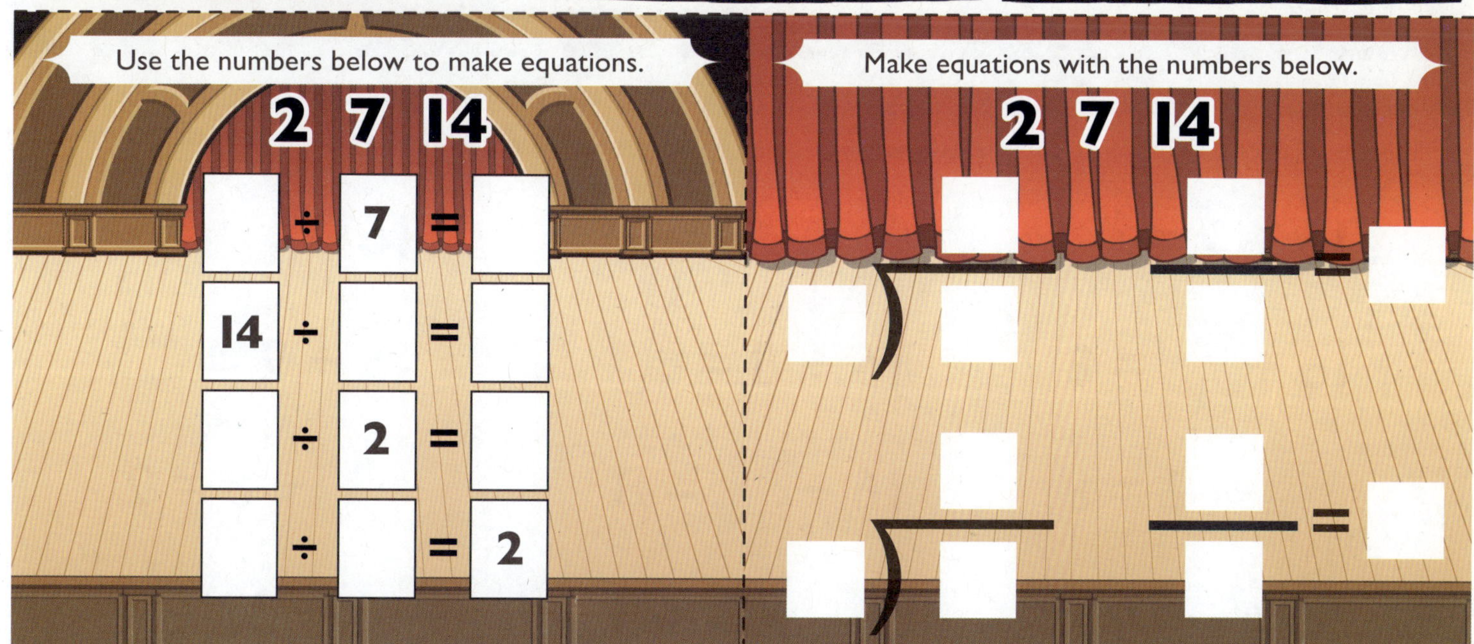

Divide the saxophones into seven even groups by circling each group.

Write the equation to match the picture. ______ ÷ ______ = ______

Fill in the Answer

Fill in the blanks to complete the equations.

$16 \div 2 = \square$ $\quad\quad$ $\square \div 8 = 2$ $\quad\quad$ $16 \div \square = 2$

$16 \div \square = 8$ $\quad\quad$ $16 \div 8 = \square$ $\quad\quad$ $16 \div 2 = \square$

$\square \div 2 = 8$ $\quad\quad$ $16 \div \square = 8$ $\quad\quad$ $\square \div 8 = 2$

$16 \div \square = 2$ $\quad\quad$ $\square \div 2 = 8$ $\quad\quad$ $16 \div 8 = \square$

$\begin{array}{r} 16 \\ \div\ \square \\ \hline 2 \end{array}$ $\quad$ $\begin{array}{r} \square \\ \div\ 2 \\ \hline 8 \end{array}$ $\quad$ $\begin{array}{r} 16 \\ \div\ 2 \\ \hline \square \end{array}$ $\quad$ $\begin{array}{r} 16 \\ \div\ 8 \\ \hline \square \end{array}$ $\quad$ $\begin{array}{r} 16 \\ \div\ \square \\ \hline 8 \end{array}$ $\quad$ $\begin{array}{r} \square \\ \div\ 8 \\ \hline 2 \end{array}$

$\begin{array}{r} \square \\ 2\overline{)16} \end{array}$ $\quad$ $\frac{16}{8} = \square$ $\quad$ $\begin{array}{r} \square \\ 8\overline{)16} \end{array}$ $\quad$ $\frac{16}{2} = \square$ $\quad$ $\begin{array}{r} \square \\ 2\overline{)16} \end{array}$

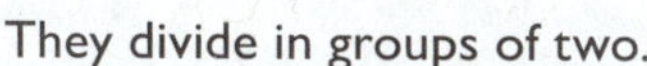

Activities

There are sixteen birds.

They divide in groups of two.

How many groups of birds are there total? ______

Make equations with the numbers below.

2 8 16

Divide the goggles into eight even groups by circling each group.

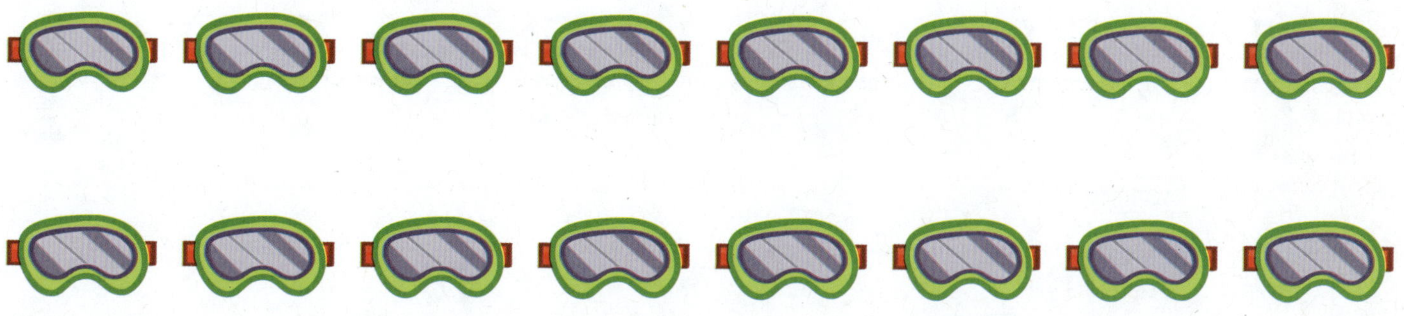

Write the equation to match the picture. ______ ÷ ______ = ______

Fill in the Answer

Fill in the blanks to complete the equations.

$18 \div \square = 9$ $\quad$ $18 \div 2 = \square$ $\quad$ $18 \div 9 = \square$

$18 \div 2 = \square$ $\quad$ $\square \div 9 = 2$ $\quad$ $\square \div 2 = 9$

$\square \div 9 = 2$ $\quad$ $18 \div \square = 9$ $\quad$ $18 \div \square = 2$

$18 \div \square = 2$ $\quad$ $18 \div 2 = \square$ $\quad$ $\square \div 9 = 2$

$\begin{array}{r} 18 \\ \div\ 2 \\ \hline \square \end{array}$ $\quad$ $\begin{array}{r} \square \\ \div\ 9 \\ \hline 2 \end{array}$ $\quad$ $\begin{array}{r} 18 \\ \div\ 9 \\ \hline \square \end{array}$ $\quad$ $\begin{array}{r} \square \\ \div\ 2 \\ \hline 9 \end{array}$ $\quad$ $\begin{array}{r} 18 \\ \div\ 2 \\ \hline \square \end{array}$ $\quad$ $\begin{array}{r} \square \\ \div\ 9 \\ \hline 2 \end{array}$

$\begin{array}{r} \square \\ 2\overline{)18} \end{array}$ $\quad$ $\dfrac{18}{2} = \square$ $\quad$ $\begin{array}{r} \square \\ 9\overline{)18} \end{array}$ $\quad$ $\dfrac{18}{9} = \square$ $\quad$ $\begin{array}{r} \square \\ 9\overline{)18} \end{array}$

Activities

There are eighteen apples.

There are two baskets.

When divided evenly, how many apples will be in each basket? ______

Use the numbers below to make equations.

2 9 18

◯ ÷ ◯ = 2

18 ÷ ◯ = ◯

◯ ÷ 2 = ◯

◯ ÷ 9 = ◯

Make equations with the numbers below.

2 9 18

Divide the butterflies into nine even groups by circling each group.

Write the equation to match the picture. ______ ÷ ______ = ______

Fill in the Answer

Fill in the blanks to complete the equations.

$20 \div 2 = \square$ $\quad 20 \div 10 = \square$ $\quad \square \div 2 = 10$

$20 \div \square = 2$ $\quad \square \div 10 = 2$ $\quad 20 \div 10 = \square$

$\square \div 2 = 10$ $\quad 20 \div 2 = \square$ $\quad \square \div 10 = 2$

$20 \div \square = 10$ $\quad 20 \div \square = 2$ $\quad 20 \div \square = 10$

$$\begin{array}{r} 20 \\ \div\ 2 \\ \hline \square \end{array} \quad \begin{array}{r} \square \\ \div\ 10 \\ \hline 2 \end{array} \quad \begin{array}{r} 20 \\ \div\ \square \\ \hline 2 \end{array} \quad \begin{array}{r} 20 \\ \div\ 10 \\ \hline \square \end{array} \quad \begin{array}{r} 20 \\ \div\ \square \\ \hline 10 \end{array} \quad \begin{array}{r} 20 \\ \div\ 2 \\ \hline \square \end{array}$$

$$2\overline{)20}\ \square \quad \frac{20}{2} = \square \quad 10\overline{)20}\ \square \quad \frac{20}{10} = \square \quad 2\overline{)20}\ \square$$

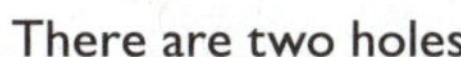

Activities

There are twenty rocks.

There are two holes.

How many rocks can be divided evenly into each hole? _____

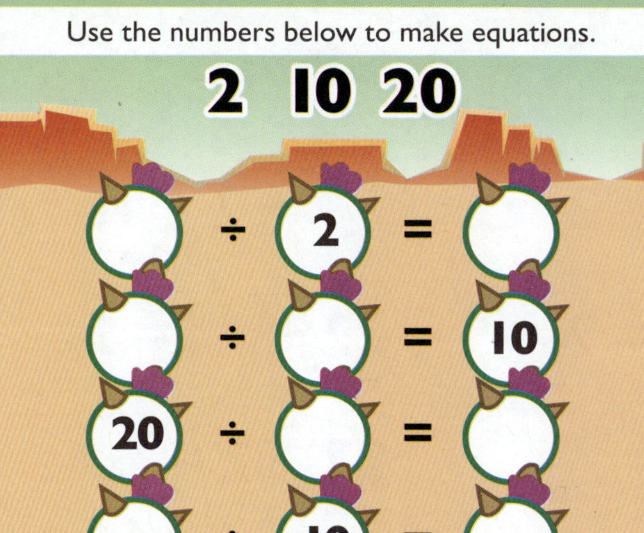

Divide the cartwheels into ten even groups by circling each group.

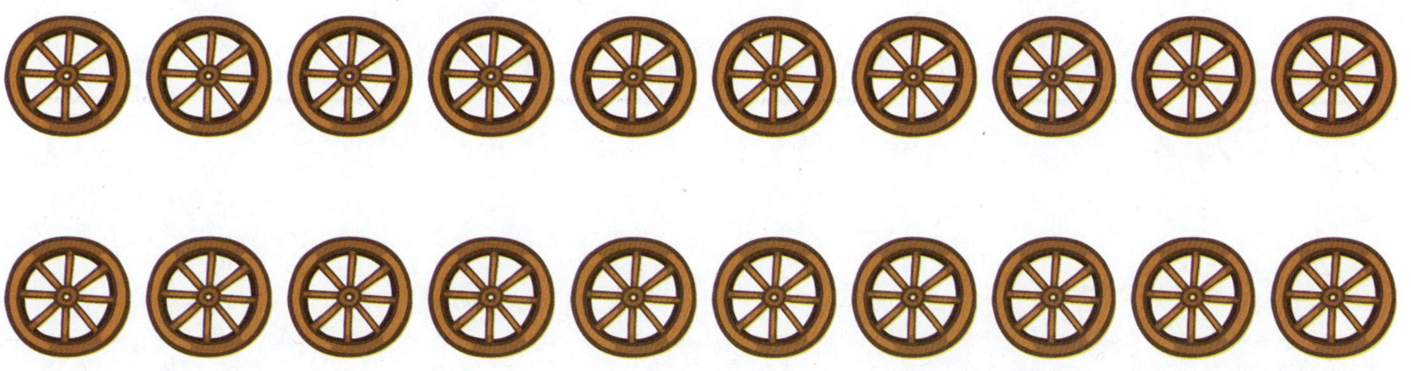

Write the equation to match the picture. _______ ÷ _______ = _______

Fill in the Answer

Fill in the blanks to complete the equations.

$22 \div 11 = \square$ $\qquad$ $22 \div \square = 2$ $\qquad$ $\square \div 2 = 11$

$\square \div 2 = 11$ $\qquad$ $\square \div 11 = 2$ $\qquad$ $22 \div \square = 2$

$22 \div \square = 11$ $\qquad$ $22 \div 11 = \square$ $\qquad$ $22 \div 2 = \square$

$22 \div 11 = \square$ $\qquad$ $22 \div \square = 11$ $\qquad$ $\square \div 11 = 2$

$$\begin{array}{r} 22 \\ \div\ \square \\ \hline 2 \end{array} \qquad \begin{array}{r} 22 \\ \div\ 11 \\ \hline \square \end{array} \qquad \begin{array}{r} \square \\ \div\ 2 \\ \hline 11 \end{array} \qquad \begin{array}{r} 22 \\ \div\ 2 \\ \hline \square \end{array} \qquad \begin{array}{r} \square \\ \div\ 11 \\ \hline 2 \end{array} \qquad \begin{array}{r} 22 \\ \div\ \square \\ \hline 11 \end{array}$$

$$2\overline{)22}\ \text{(quotient: } \square) \qquad \frac{22}{2} = \square \qquad 11\overline{)22}\ \text{(quotient: } \square) \qquad \frac{22}{11} = \square \qquad 2\overline{)22}\ \text{(quotient: } \square)$$

Activities

There are twenty-two bells.

There are two trees.

How many bells will hang from each tree when they are divided evenly? _____

Use the numbers below to make equations.

2 11 22

☐ ÷ ☐ = 2

☐ ÷ ☐ = 11

☐ ÷ 2 = ☐

22 ÷ ☐ = ☐

Make equations with the numbers below.

2 11 22

Divide the bells into eleven even groups by circling each group.

Write the equation to match the picture. ______ ÷ ______ = ______

Fill in the Answer

Fill in the blanks to complete the equations.

$24 \div \square = 12$	$\square \div 12 = 2$	$\square \div 2 = 12$
$24 \div 2 = \square$	$24 \div 12 = \square$	$\square \div 12 = 2$
$\square \div 2 = 12$	$24 \div \square = 2$	$\square \div 12 = 2$
$24 \div \square = 2$	$24 \div 2 = \square$	$24 \div \square = 12$

$\begin{array}{r} \square \\ \div\ 2 \\ \hline 12 \end{array}$ $\begin{array}{r} 24 \\ \div\ 2 \\ \hline \square \end{array}$ $\begin{array}{r} 24 \\ \div\ \square \\ \hline 12 \end{array}$ $\begin{array}{r} \square \\ \div\ 12 \\ \hline 2 \end{array}$ $\begin{array}{r} 24 \\ \div\ 12 \\ \hline \square \end{array}$ $\begin{array}{r} 24 \\ \div\ \square \\ \hline 2 \end{array}$

$\begin{array}{r} \square \\ 2\overline{)24} \end{array}$ $\frac{24}{2} = \square$ $\begin{array}{r} \square \\ 12\overline{)24} \end{array}$ $\frac{24}{12} = \square$ $\begin{array}{r} \square \\ 2\overline{)24} \end{array}$

Activities

There are twenty-four cookies.

There are two trays.

When divided evenly, how many cookies will be on each tray? ______

Use the numbers below to make equations.

2 12 24

☐ ÷ 2 = ☐

☐ ÷ ☐ = 2

☐ ÷ 12 = ☐

24 ÷ ☐ = ☐

Make equations with the numbers below.

2 12 24

☐)☐ with ☐ above ☐/☐ = ☐

☐)☐ with ☐ above ☐/☐ = ☐

Divide the cookies into twelve even groups by circling each group.

Write the equation to match the picture. ______ ÷ ______ = ______

Fill in the Answer

Fill in the blanks to complete the equations.

$\square \div 3 = 3$	$9 \div 3 = \square$	$\square \div 3 = 3$
$9 \div 3 = \square$	$9 \div \square = 3$	$9 \div 3 = \square$
$9 \div \square = 3$	$\square \div 3 = 3$	$\square \div 3 = 3$
$9 \div \square = 3$	$9 \div 3 = \square$	$9 \div \square = 3$

$$\begin{array}{r} 9 \\ \div\ \square \\ \hline 3 \end{array} \qquad \begin{array}{r} 9 \\ \div\ 3 \\ \hline \square \end{array} \qquad \begin{array}{r} \square \\ \div\ 3 \\ \hline 3 \end{array} \qquad \begin{array}{r} 9 \\ \div\ 3 \\ \hline \square \end{array} \qquad \begin{array}{r} 9 \\ \div\ \square \\ \hline 3 \end{array} \qquad \begin{array}{r} 9 \\ \div\ \square \\ \hline 3 \end{array}$$

$$\begin{array}{r} \square \\ 3\overline{)9} \end{array} \qquad \frac{9}{3} = \square \qquad \begin{array}{r} \square \\ 3\overline{)9} \end{array} \qquad \frac{9}{3} = \square \qquad \begin{array}{r} \square \\ 3\overline{)9} \end{array}$$

Activities

There are nine shells.

There are three buckets.

How many shells will be in each bucket if they are divided into equal groups? ______

Use the numbers below to make equations.

3 3 9

9 ÷ ___ = ___

___ ÷ ___ = 3

___ ÷ 3 = ___

___ ÷ 3 = ___

Make equations with the numbers below.

3 3 9

Divide the shells into three even groups by circling each group.

Write the equation to match the picture. ______ ÷ ______ = ______

Fill in the Answer

Fill in the blanks to complete the equations.

12 ÷ 3 = ___	12 ÷ ___ = 4	12 ÷ 3 = ___
12 ÷ ___ = 3	___ ÷ 4 = 3	12 ÷ ___ = 4
___ ÷ 3 = 4	12 ÷ ___ = 3	___ ÷ 4 = 3
12 ÷ 4 = ___	___ ÷ 3 = 4	12 ÷ 4 = ___

12	12	___	12	___	12
÷ ___	÷ 3	÷ 3	÷ 4	÷ 4	÷ ___
3	___	4	___	3	4

$3\overline{)12}$ ___ $\frac{12}{4} =$ ___ $4\overline{)12}$ ___ $\frac{12}{3} =$ ___ $4\overline{)12}$ ___

Activities

There are twelve bottles of hair spray.

There are three bags

When divided evenly, how many bottles will be in each bag? ______

Use the numbers below to make equations.

3 4 12

___ ÷ ___ = 4

12 ÷ ___ = ___

___ ÷ 4 = ___

___ ÷ ___ = 3

Make equations with the numbers below.

3 4 12

Divide the spray bottles into four even groups by circling each group.

Write the equation to match the picture. ______ ÷ ______ = ______

Fill in the Answer

Fill in the blanks to complete the equations.

$15 \div 3 = \square$ $\square \div 5 = 3$ $\square \div 5 = 3$

$15 \div 5 = \square$ $15 \div \square = 5$ $15 \div \square = 5$

$15 \div \square = 3$ $\square \div 3 = 5$ $15 \div 3 = \square$

$\square \div 3 = 5$ $15 \div \square = 3$ $15 \div 5 = \square$

$$\begin{array}{r}\square\\ \div\ 5\\ \hline 3\end{array} \quad \begin{array}{r}15\\ \div\ 3\\ \hline \square\end{array} \quad \begin{array}{r}15\\ \div\ \square\\ \hline 3\end{array} \quad \begin{array}{r}\square\\ \div\ 3\\ \hline 5\end{array} \quad \begin{array}{r}15\\ \div\ \square\\ \hline 5\end{array} \quad \begin{array}{r}15\\ \div\ 5\\ \hline \square\end{array}$$

$$3\overline{)15}\ (\square) \quad \frac{15}{3} = \square \quad 5\overline{)15}\ (\square) \quad \frac{15}{5} = \square \quad 3\overline{)15}\ (\square)$$

Activities

There are fifteen gems.

There are five carts.

When divided evenly, how many gems will be in each cart? ______

Use the numbers below to make equations.

3 5 15

___ ÷ 3 = ___

___ ÷ ___ = 5

___ ÷ 5 = ___

15 ÷ ___ = ___

Divide the gems into five even groups by circling each group.

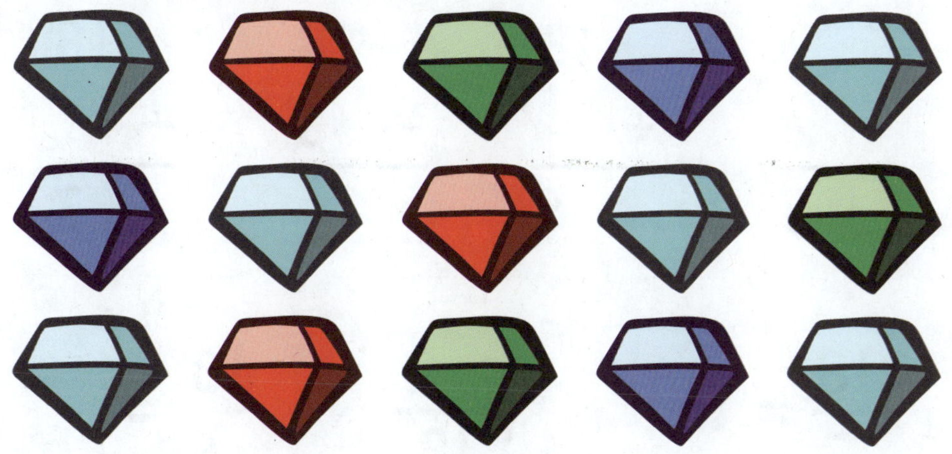

Write the equation to match the picture. ______ ÷ ______ = ______

Fill in the Answer

Fill in the blanks to complete the equations.

$\square \div 3 = 6$ $\quad$ $18 \div 3 = \square$ $\quad$ $18 \div 6 = \square$

$18 \div 3 = \square$ $\quad$ $\square \div 6 = 3$ $\quad$ $\square \div 6 = 3$

$18 \div \square = 6$ $\quad$ $18 \div \square = 3$ $\quad$ $18 \div \square = 6$

$18 \div 6 = \square$ $\quad$ $\square \div 3 = 6$ $\quad$ $18 \div \square = 3$

$$\begin{array}{r} 18 \\ \div\ 3 \\ \hline \square \end{array} \quad \begin{array}{r} \square \\ \div\ 6 \\ \hline 3 \end{array} \quad \begin{array}{r} 18 \\ \div\ \square \\ \hline 6 \end{array} \quad \begin{array}{r} 18 \\ \div\ 6 \\ \hline \square \end{array} \quad \begin{array}{r} 18 \\ \div\ \square \\ \hline 3 \end{array} \quad \begin{array}{r} 18 \\ \div\ 3 \\ \hline \square \end{array}$$

$$3\overline{)\,18}\ \ (\square) \quad \frac{18}{3} = \square \quad 6\overline{)\,18}\ \ (\square) \quad \frac{18}{6} = \square \quad 3\overline{)\,18}\ \ (\square)$$

Activities

There are eighteen flowers.

There are six vases.

How many flowers will be in each vase if they are divided into equal groups? _____

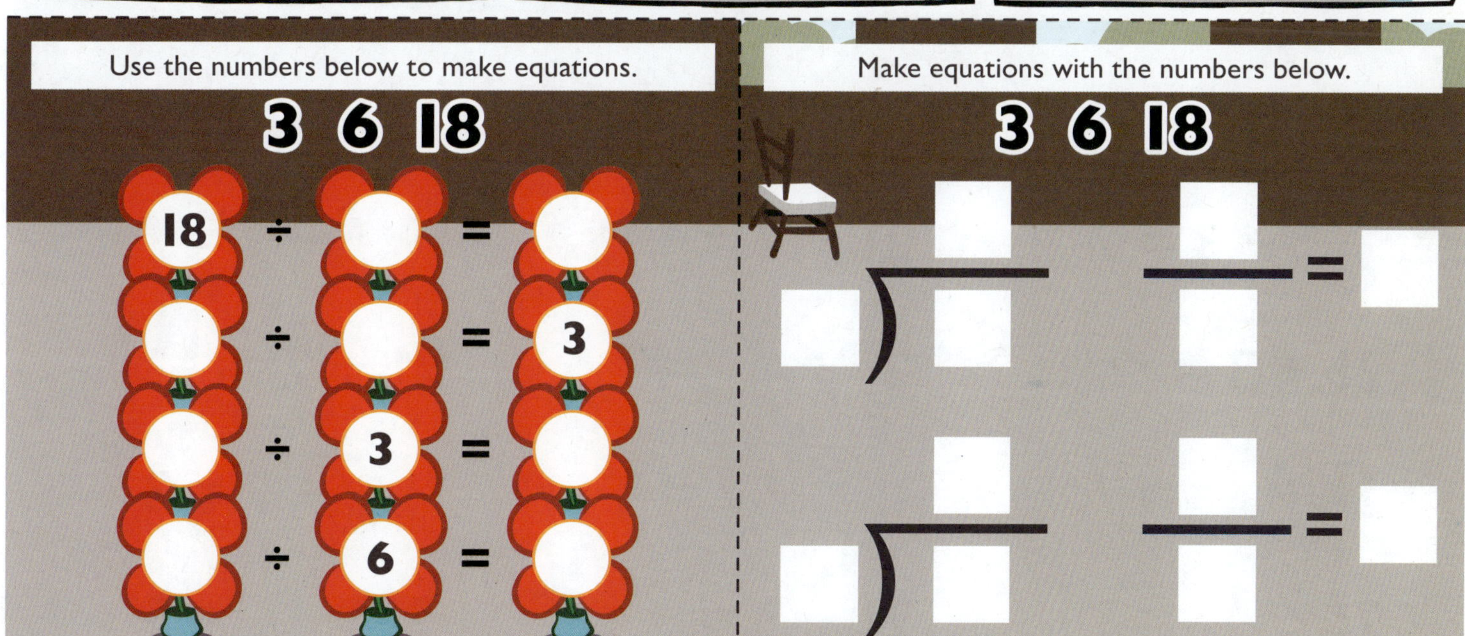

Divide the flowers into three even groups by circling each group.

Write the equation to match the picture. _______ ÷ _______ = _______

Fill in the Answer

Fill in the blanks to complete the equations.

☐ ÷ 3 = 7	21 ÷ 7 = ☐	☐ ÷ 3 = 7
21 ÷ 3 = ☐	21 ÷ ☐ = 7	21 ÷ 7 = ☐
21 ÷ ☐ = 7	☐ ÷ 7 = 3	☐ ÷ 7 = 3
21 ÷ ☐ = 3	21 ÷ 3 = ☐	21 ÷ ☐ = 3

$$\begin{array}{r} 21 \\ \div\ \square \\ \hline 3 \end{array} \quad \begin{array}{r} 21 \\ \div\ 3 \\ \hline \square \end{array} \quad \begin{array}{r} \square \\ \div\ 7 \\ \hline 3 \end{array} \quad \begin{array}{r} 21 \\ \div\ 7 \\ \hline \square \end{array} \quad \begin{array}{r} \square \\ \div\ 3 \\ \hline 7 \end{array} \quad \begin{array}{r} 21 \\ \div\ \square \\ \hline 7 \end{array}$$

$$3\overline{)21}^{\ \square} \quad \frac{21}{3} = \square \quad 7\overline{)21}^{\ \square} \quad \frac{21}{7} = \square \quad 3\overline{)21}^{\ \square}$$

Activities

There are twenty-one sticks. The sticks are in groups of seven. How many groups of sticks are there? ___

Use the numbers below to make equations.

3 7 21

21 ÷ ___ = ___

___ ÷ ___ = 3

___ ÷ 3 = ___

___ ÷ 7 = ___

Make equations with the numbers below.

3 7 21

Divide the sleds into three even groups by circling each group.

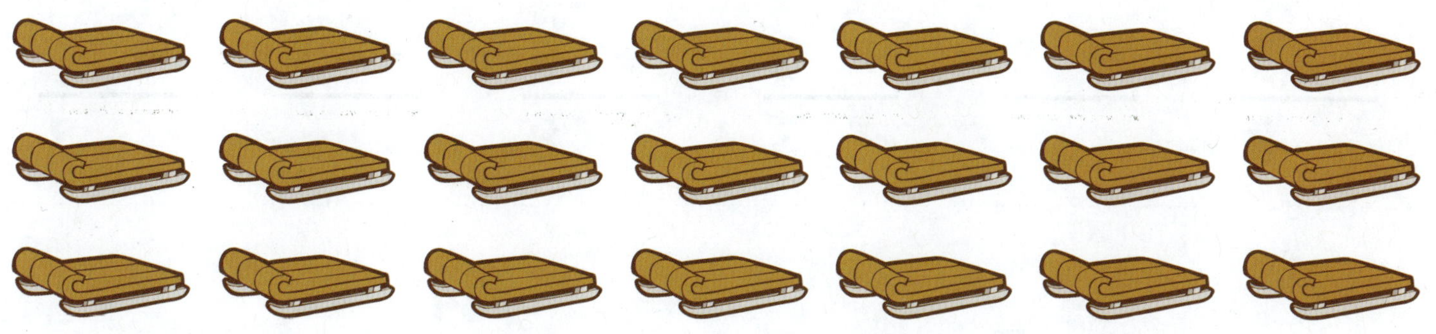

Write the equation to match the picture. ______ ÷ ______ = ______

Fill in the Answer

Fill in the blanks to complete the equations.

$24 \div \square = 3$ $\quad$ $24 \div 8 = \square$ $\quad$ $24 \div \square = 3$

$\square \div 3 = 8$ $\quad$ $\square \div 8 = 3$ $\quad$ $24 \div 8 = \square$

$24 \div 3 = \square$ $\quad$ $24 \div 8 = \square$ $\quad$ $24 \div \square = 8$

$\square \div 8 = 3$ $\quad$ $24 \div \square = 8$ $\quad$ $\square \div 8 = 3$

$$\begin{array}{r}\square\\ \div\ 3\\ \hline 8\end{array}\quad \begin{array}{r}24\\ \div\ 8\\ \hline \square\end{array}\quad \begin{array}{r}\square\\ \div\ 8\\ \hline 3\end{array}\quad \begin{array}{r}\square\\ \div\ 3\\ \hline 8\end{array}\quad \begin{array}{r}24\\ \div\ 3\\ \hline \square\end{array}\quad \begin{array}{r}24\\ \div\ \square\\ \hline 3\end{array}$$

$$3\overline{)24}^{\,\square}\qquad \frac{24}{8} = \square \qquad 8\overline{)24}^{\,\square}\qquad \frac{24}{3} = \square \qquad 3\overline{)24}^{\,\square}$$

Activities

There are twenty-four snowballs.

There are eight shovels.

How many snowballs will be in each shovel when divided evenly? _____

Use the numbers below to make equations.

3 8 24

24 ÷ ___ = ___

___ ÷ ___ = 3

___ ÷ 8 = ___

___ ÷ ___ = 8

Make equations with the numbers below.

3 8 24

Divide the snowmen into three even groups by circling each group.

Write the equation to match the picture. ______ ÷ ______ = ______

Fill in the Answer

Fill in the blanks to complete the equations.

$27 \div \square = 9$ $\quad$ $27 \div 9 = \square$ $\quad$ $27 \div 3 = \square$

$\square \div 9 = 3$ $\quad$ $27 \div \square = 9$ $\quad$ $27 \div \square = 3$

$27 \div \square = 9$ $\quad$ $\square \div 9 = 3$ $\quad$ $\square \div 3 = 9$

$\square \div 9 = 3$ $\quad$ $27 \div 9 = \square$ $\quad$ $27 \div 3 = \square$

$$\begin{array}{r} 27 \\ \div\ 3 \\ \hline \square \end{array} \quad \begin{array}{r} 27 \\ \div\ \square \\ \hline 9 \end{array} \quad \begin{array}{r} \square \\ \div\ 3 \\ \hline 9 \end{array} \quad \begin{array}{r} \square \\ \div\ 9 \\ \hline 3 \end{array} \quad \begin{array}{r} 27 \\ \div\ 9 \\ \hline \square \end{array} \quad \begin{array}{r} \square \\ \div\ 3 \\ \hline 9 \end{array}$$

$$\begin{array}{r} \square \\ 9\overline{)27} \end{array} \quad \frac{27}{3} = \square \quad \begin{array}{r} \square \\ 3\overline{)27} \end{array} \quad \frac{27}{9} = \square \quad \begin{array}{r} \square \\ 9\overline{)27} \end{array}$$

Activities

There are twenty-seven balls of yarn.

There are nine baskets.

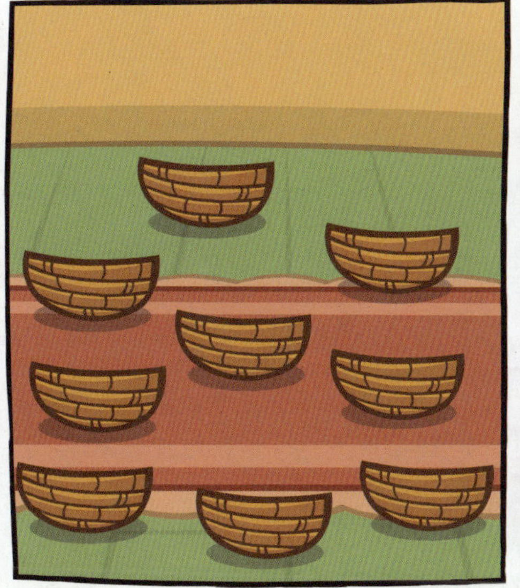

How many balls of yarn are there in each basket when divided evenly? ______

Use the numbers below to make equations.

3 9 27

( ) ÷ ( ) = (9)

(27) ÷ ( ) = ( )

( ) ÷ (3) = ( )

( ) ÷ (9) = ( )

Make equations with the numbers below.

3 9 27

Divide the tennis balls into three even groups by circling each group.

Write the equation to match the picture. ______ ÷ ______ = ______

Fill in the Answer

Fill in the blanks to complete the equations.

$30 \div 3 = \square$ $30 \div \square = 10$ $30 \div \square = 3$

$30 \div \square = 3$ $\square \div 3 = 10$ $30 \div 10 = \square$

$\square \div 3 = 10$ $30 \div 3 = \square$ $\square \div 3 = 10$

$30 \div 10 = \square$ $\square \div 10 = 3$ $30 \div \square = 10$

$\begin{array}{r} 30 \\ \div \square \\ \hline 3 \end{array}$ $\begin{array}{r} 30 \\ \div 10 \\ \hline \square \end{array}$ $\begin{array}{r} \square \\ \div 10 \\ \hline 3 \end{array}$ $\begin{array}{r} 30 \\ \div 3 \\ \hline \square \end{array}$ $\begin{array}{r} \square \\ \div 3 \\ \hline 10 \end{array}$ $\begin{array}{r} 30 \\ \div \square \\ \hline 10 \end{array}$

$3\overline{)30}$ (answer: $\square$) $\frac{30}{3} = \square$ $10\overline{)30}$ (answer: $\square$) $\frac{30}{10} = \square$ $3\overline{)30}$ (answer: $\square$)

Activities

There are thirty apples.

There are three bags.

When divided evenly, how many apples will be in each bag? _____

Use the numbers below to make equations.

3 10 30

___ ÷ ___ = 3

30 ÷ ___ = ___

___ ÷ 3 = ___

___ ÷ 10 = ___

Make equations with the numbers below.

3 10 30

Divide the cans into ten even groups by circling each group.

Write the equation to match the picture. ______ ÷ ______ = ______

Fill in the Answer

Fill in the blanks to complete the equations.

$33 \div \square = 3$ $\square \div 11 = 3$ $33 \div 3 = \square$

$33 \div 3 = \square$ $33 \div \square = 3$ $33 \div \square = 11$

$33 \div \square = 3$ $\square \div 3 = 11$ $\square \div 11 = 3$

$\square \div 3 = 11$ $33 \div 11 = \square$ $33 \div 11 = \square$

$\begin{array}{r} 33 \\ \div\ 3 \\ \hline \square \end{array}$ $\begin{array}{r} 33 \\ \div\ \square \\ \hline 11 \end{array}$ $\begin{array}{r} 33 \\ \div\ 3 \\ \hline \square \end{array}$ $\begin{array}{r} \square \\ \div\ 3 \\ \hline 11 \end{array}$ $\begin{array}{r} \square \\ \div\ 3 \\ \hline 11 \end{array}$ $\begin{array}{r} 33 \\ \div\ 3 \\ \hline \square \end{array}$

$3\overline{)33}$ (answer: $\square$) $\frac{33}{3} = \square$ $3\overline{)33}$ (answer: $\square$) $\frac{33}{11} = \square$ $11\overline{)33}$ (answer: $\square$)

Activities

There are thirty bubbles.

There are eleven sea squirts.

When divided evenly, how many bubbles does each sea squirt blow? ______

Use the numbers below to make equations.

3 11 33

$\bigcirc \div \bigcirc = 3$

$\bigcirc \div 3 = \bigcirc$

$\bigcirc \div 11 = \bigcirc$

$33 \div \bigcirc = \bigcirc$

Make equations with the numbers below.

Divide the bubbles into three even groups by circling each group.

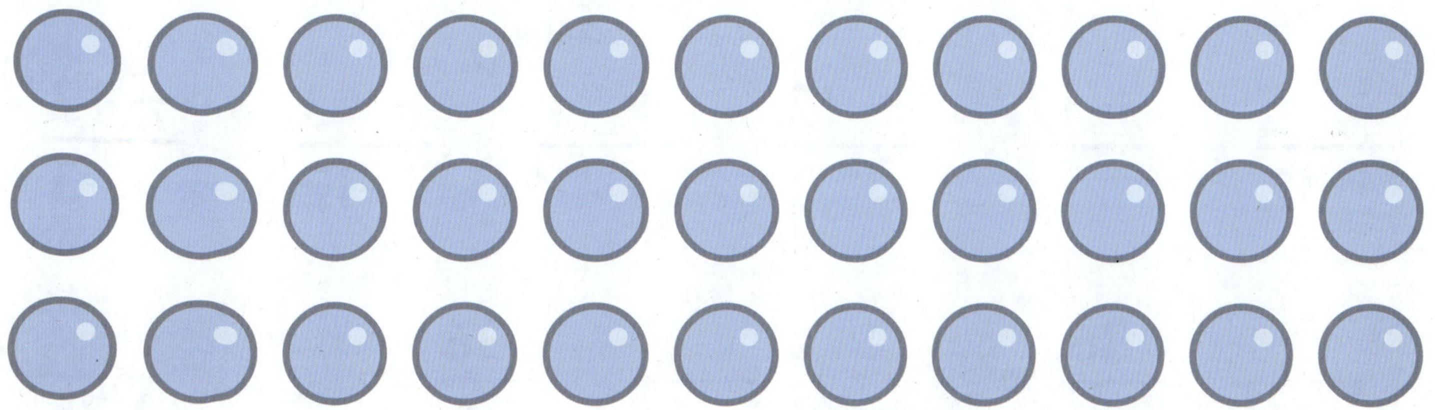

Write the equation to match the picture. ______ ÷ ______ = ______

Fill in the Answer

Fill in the blanks to complete the equations.

$\square \div 3 = 12$	$36 \div \square = 12$	$36 \div 12 = \square$
$36 \div \square = 3$	$\square \div 3 = 12$	$36 \div \square = 12$
$36 \div 12 = \square$	$36 \div 3 = \square$	$\square \div 12 = 3$
$\square \div 12 = 3$	$36 \div \square = 3$	$36 \div 3 = \square$

$$\begin{array}{r}36\\ \div\ \square\\ \hline 12\end{array}\qquad\begin{array}{r}36\\ \div\ 12\\ \hline \square\end{array}\qquad\begin{array}{r}\square\\ \div\ 12\\ \hline 3\end{array}\qquad\begin{array}{r}36\\ \div\ 3\\ \hline \square\end{array}\qquad\begin{array}{r}\square\\ \div\ 3\\ \hline 12\end{array}\qquad\begin{array}{r}\square\\ \div\ 12\\ \hline 3\end{array}$$

$$\begin{array}{r}\square\\ 3\overline{)36}\end{array}\qquad \frac{36}{3} = \square \qquad \begin{array}{r}\square\\ 12\overline{)36}\end{array}\qquad \frac{36}{12} = \square \qquad \begin{array}{r}\square\\ 3\overline{)36}\end{array}$$

Activities

There are thirty-six rocks.

There are three buckets.

How many rocks does each bucket hold when divided evenly? _____

Use the numbers below to make equations.

Make equations with the numbers below.

3 12 36

Divide the harps into twelve even groups by circling each group.

Write the equation to match the picture. ______÷______=______

Fill in the Answer

Fill in the blanks to complete the equations.

$16 \div \square = 4$	$16 \div 4 = \square$	$\square \div 4 = 4$
$\square \div 4 = 4$	$16 \div \square = 4$	$16 \div 4 = \square$
$16 \div 4 = \square$	$\square \div 4 = 4$	$\square \div 4 = 4$
$16 \div \square = 4$	$16 \div 4 = \square$	$16 \div \square = 4$

$$\begin{array}{r} 16 \\ \div\ 4 \\ \hline \square \end{array} \quad \begin{array}{r} \square \\ \div\ 4 \\ \hline 4 \end{array} \quad \begin{array}{r} \square \\ \div\ 4 \\ \hline 4 \end{array} \quad \begin{array}{r} \square \\ \div\ 4 \\ \hline 4 \end{array} \quad \begin{array}{r} 16 \\ \div\ \square \\ \hline 4 \end{array} \quad \begin{array}{r} 16 \\ \div\ 4 \\ \hline \square \end{array}$$

$$\begin{array}{r} \square \\ 4\overline{)16} \end{array} \quad \frac{16}{4} = \square \quad \begin{array}{r} \square \\ 4\overline{)16} \end{array} \quad \frac{16}{4} = \square \quad \begin{array}{r} \square \\ 4\overline{)16} \end{array}$$

Activities

There are sixteen coins. Each game costs four coins to play. How many games can we play? ____

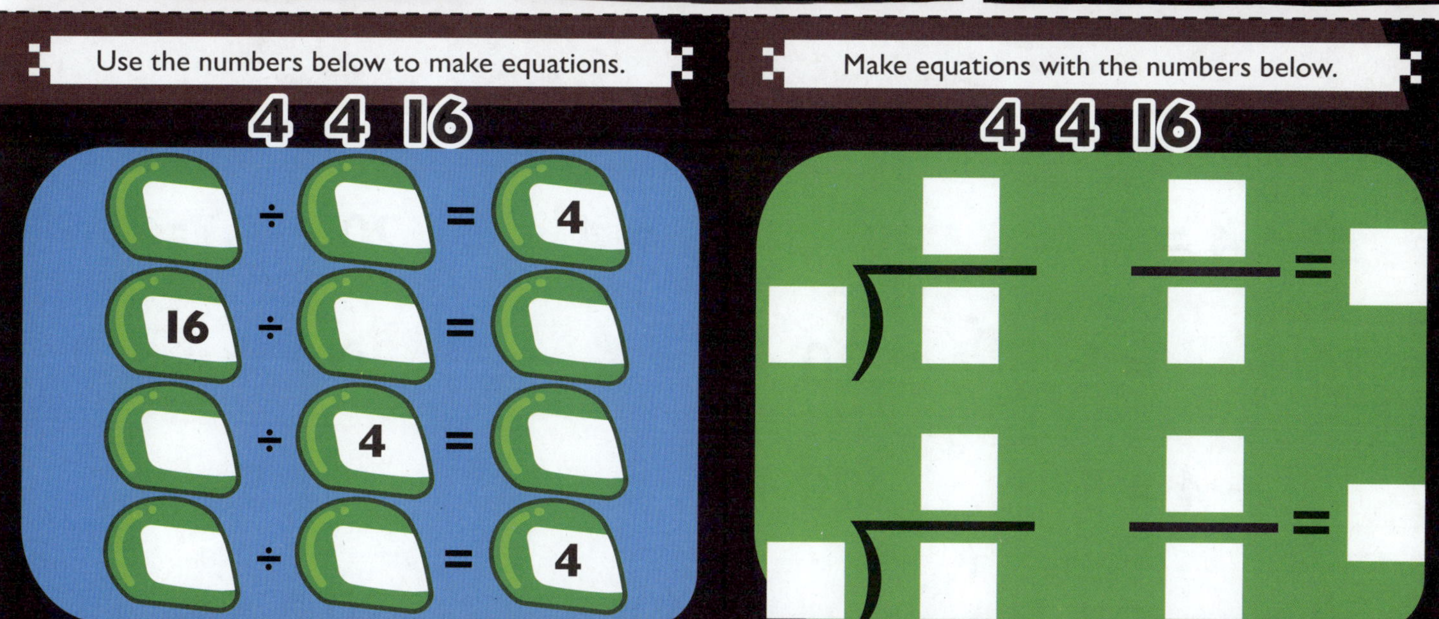

Divide the helmets into four even groups by circling each group.

Write the equation to match the picture. ______ ÷ ______ = ______

Fill in the Answer

Fill in the blanks to complete the equations.

$20 \div 4 = \square$	$\square \div 5 = 4$	$20 \div \square = 4$
$20 \div \square = 5$	$20 \div 5 = \square$	$20 \div 4 = \square$
$\square \div 4 = 5$	$20 \div \square = 5$	$\square \div 5 = 4$
$20 \div \square = 4$	$\square \div 4 = 5$	$20 \div 5 = \square$

$\begin{array}{r} 20 \\ \div\ \square \\ \hline 4 \end{array}$ $\begin{array}{r} \square \\ \div\ 4 \\ \hline 5 \end{array}$ $\begin{array}{r} 20 \\ \div\ 4 \\ \hline \square \end{array}$ $\begin{array}{r} 20 \\ \div\ 5 \\ \hline \square \end{array}$ $\begin{array}{r} 20 \\ \div\ \square \\ \hline 5 \end{array}$ $\begin{array}{r} \square \\ \div\ 5 \\ \hline 4 \end{array}$

$4\overset{\square}{\overline{)20}}$ $\frac{20}{5} = \square$ $4\overset{\square}{\overline{)20}}$ $\frac{20}{4} = \square$ $5\overset{\square}{\overline{)20}}$

Activities

There are twenty carrots.

There are four spears.

How many carrots can we put on each spear in equal amounts? ______

Use the numbers below to make equations.

4 5 20

	÷		=	5
20	÷		=	
	÷	4	=	
	÷		=	4

Make equations with the numbers below.

4 5 20

$\square \overline{)\square}$ with quotient $\square$ $\frac{\square}{\square} = \square$

$\square \overline{)\square}$ with quotient $\square$ $\frac{\square}{\square} = \square$

Divide the tomatoes into five even groups by circling each group.

Write the equation to match the picture. ______ ÷ ______ = ______

Fill in the Answer

Fill in the blanks to complete the equations.

$24 \div \square = 6$ $24 \div 4 = \square$ $24 \div 6 = \square$

$24 \div 4 = \square$ $\square \div 6 = 4$ $\square \div 4 = 6$

$\square \div 4 = 6$ $24 \div \square = 4$ $24 \div \square = 6$

$24 \div \square = 4$ $24 \div 4 = \square$ $\square \div 6 = 4$

$$\begin{array}{r} 24 \\ \div\ 4 \\ \hline \square \end{array} \quad \begin{array}{r} \square \\ \div\ 6 \\ \hline 4 \end{array} \quad \begin{array}{r} 24 \\ \div\ 6 \\ \hline \square \end{array} \quad \begin{array}{r} \square \\ \div\ 4 \\ \hline 6 \end{array} \quad \begin{array}{r} 24 \\ \div\ 4 \\ \hline \square \end{array} \quad \begin{array}{r} \square \\ \div\ 6 \\ \hline 4 \end{array}$$

$$4\overline{)24} = \square \quad \frac{24}{4} = \square \quad 6\overline{)24} = \square \quad \frac{24}{6} = \square \quad 4\overline{)24} = \square$$

Activities

There are twenty-four marbles.

There are four bags.

How many marbles can be divided equally into each bag? ______

Use the numbers below to make equations.

4 6 24

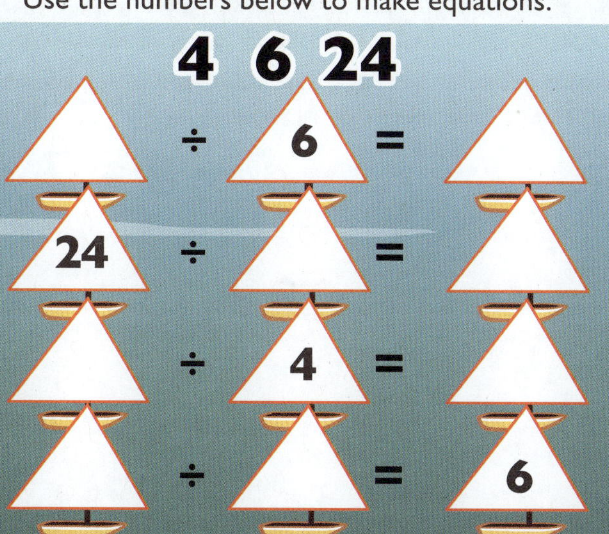

Make equations with the numbers below.

4 6 24

Divide the joy sticks into six even groups by circling each group.

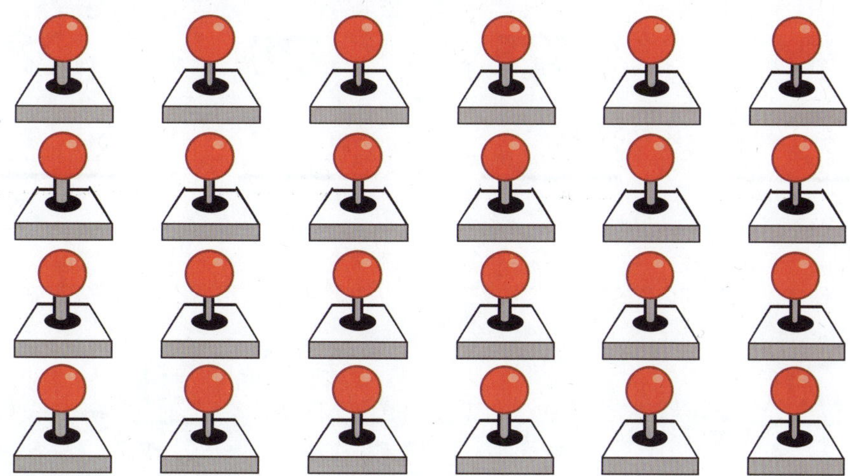

Write the equation to match the picture. ______ ÷ ______ = ______

Fill in the Answer

Fill in the blanks to complete the equations.

$28 \div 4 = \square$ $\quad$ $28 \div 7 = \square$ $\quad$ $\square \div 4 = 7$

$28 \div \square = 4$ $\quad$ $\square \div 7 = 4$ $\quad$ $28 \div 7 = \square$

$\square \div 4 = 7$ $\quad$ $28 \div 4 = \square$ $\quad$ $\square \div 7 = 4$

$28 \div \square = 7$ $\quad$ $28 \div \square = 4$ $\quad$ $28 \div \square = 7$

$\begin{array}{r} 28 \\ \div\ 4 \\ \hline \square \end{array}$ $\quad$ $\begin{array}{r} \square \\ \div\ 7 \\ \hline 4 \end{array}$ $\quad$ $\begin{array}{r} 28 \\ \div\ \square \\ \hline 4 \end{array}$ $\quad$ $\begin{array}{r} 28 \\ \div\ 7 \\ \hline \square \end{array}$ $\quad$ $\begin{array}{r} 28 \\ \div\ \square \\ \hline 7 \end{array}$ $\quad$ $\begin{array}{r} 28 \\ \div\ 4 \\ \hline \square \end{array}$

$4\overline{)28}$ with $\square$ above $\quad$ $\frac{28}{4} = \square$ $\quad$ $7\overline{)28}$ with $\square$ above $\quad$ $\frac{28}{7} = \square$ $\quad$ $4\overline{)28}$ with $\square$ above

Activities

She has twenty-eight coins.

She pays seven coins for a purse.

How many purses can she buy with her coins? ______

Use the numbers below to make equations.

4 7 28

___	÷	___	=	4
28	÷	___	=	___
___	÷	7	=	___
___	÷	4	=	___

Make equations with the numbers below.

4 7 28

Divide the purses into four even groups by circling each group.

Write the equation to match the picture. ______ ÷ ______ = ______

Fill in the Answer

Fill in the blanks to complete the equations.

$32 \div 8 = \square$	$32 \div \square = 8$	$\square \div 4 = 8$
$\square \div 8 = 4$	$\square \div 4 = 8$	$32 \div \square = 8$
$32 \div \square = 4$	$32 \div 8 = \square$	$32 \div 4 = \square$
$32 \div 8 = \square$	$32 \div \square = 4$	$\square \div 4 = 8$

$$\begin{array}{r} 32 \\ \div\ \square \\ \hline 4 \end{array} \qquad \begin{array}{r} 32 \\ \div\ 8 \\ \hline \square \end{array} \qquad \begin{array}{r} \square \\ \div\ 4 \\ \hline 8 \end{array} \qquad \begin{array}{r} 32 \\ \div\ 4 \\ \hline \square \end{array} \qquad \begin{array}{r} \square \\ \div\ 8 \\ \hline 4 \end{array} \qquad \begin{array}{r} 32 \\ \div\ \square \\ \hline 8 \end{array}$$

$$\begin{array}{r} \square \\ 4\overline{)32} \end{array} \qquad \frac{32}{4} = \square \qquad \begin{array}{r} \square \\ 8\overline{)32} \end{array} \qquad \frac{32}{8} = \square \qquad \begin{array}{r} \square \\ 4\overline{)32} \end{array}$$

Activities

There are thirty-two carrots.

Eight carrots fit in a pot.

How many pots will we need to fit all of the carrots? _____

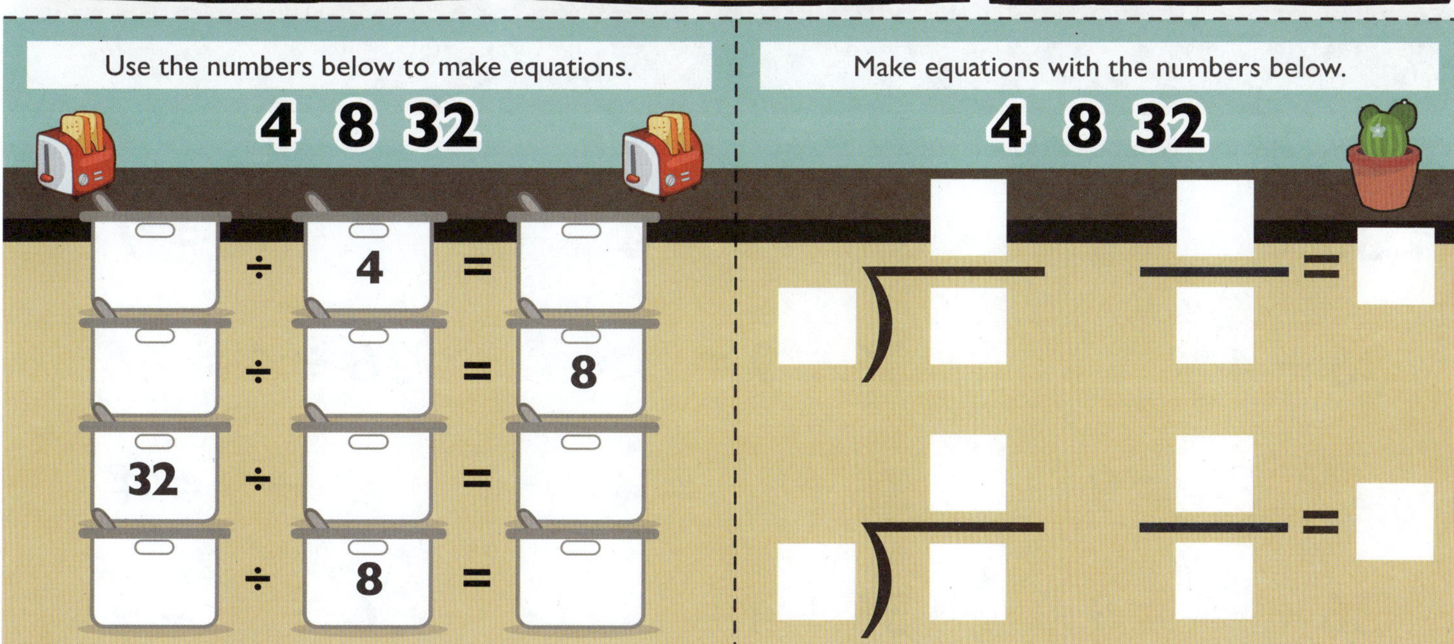

Divide the toasters into four even groups by circling each group.

Write the equation to match the picture. ______ ÷ ______ = ______

Fill in the Answer

Fill in the blanks to complete the equations.

$36 \div \square = 4$	$\square \div 9 = 4$	$\square \div 4 = 9$
$36 \div 4 = \square$	$36 \div 9 = \square$	$36 \div 4 = \square$
$\square \div 4 = 9$	$36 \div \square = 4$	$\square \div 9 = 4$
$36 \div \square = 9$	$36 \div 4 = \square$	$36 \div \square = 9$

$$\begin{array}{r}\square\\ \div\ 4\\ \hline 9\end{array}\quad \begin{array}{r}36\\ \div\ 4\\ \hline \square\end{array}\quad \begin{array}{r}36\\ \div\ \square\\ \hline 9\end{array}\quad \begin{array}{r}\square\\ \div\ 9\\ \hline 4\end{array}\quad \begin{array}{r}36\\ \div\ 9\\ \hline \square\end{array}\quad \begin{array}{r}36\\ \div\ \square\\ \hline 4\end{array}$$

$$\begin{array}{r}\square\\ 4\overline{)36}\end{array}\quad \frac{36}{4} = \square \quad \begin{array}{r}\square\\ 9\overline{)36}\end{array}\quad \frac{36}{9} = \square \quad \begin{array}{r}\square\\ 4\overline{)36}\end{array}$$

Activities

There are thirty-six coins.

A treasure chest carries nine coins.

How many chests do we need to hold all of the coins? ______

Use the numbers below to make equations.

4 9 36

$\bigcirc \div \bigcirc = 9$

$\bigcirc \div \bigcirc = 4$

$\bigcirc \div 4 = \bigcirc$

$36 \div \bigcirc = \bigcirc$

Make equations with the numbers below.

Divide the treasure chest into four even groups by circling each group.

Write the equation to match the picture. ______ ÷ ______ = ______

Fill in the Answer

Fill in the blanks to complete the equations.

$\square \div 4 = 10$ $\qquad 40 \div 10 = \square$ $\qquad \square \div 4 = 10$

$40 \div 4 = \square$ $\qquad 40 \div \square = 4$ $\qquad 40 \div 10 = \square$

$40 \div \square = 4$ $\qquad \square \div 10 = 4$ $\qquad \square \div 10 = 4$

$40 \div \square = 10$ $\qquad 40 \div 4 = \square$ $\qquad 40 \div \square = 10$

$\begin{array}{r} 40 \\ \div\ \square \\ \hline 10 \end{array}$ $\qquad \begin{array}{r} 40 \\ \div\ 4 \\ \hline \square \end{array}$ $\qquad \begin{array}{r} \square \\ \div\ 10 \\ \hline 4 \end{array}$ $\qquad \begin{array}{r} 40 \\ \div\ 10 \\ \hline \square \end{array}$ $\qquad \begin{array}{r} 40 \\ \div\ \square \\ \hline 4 \end{array}$ $\qquad \begin{array}{r} 40 \\ \div\ \square \\ \hline 10 \end{array}$

$4\overline{)40}$ quotient $\square$ $\qquad \frac{40}{4} = \square$ $\qquad 10\overline{)40}$ quotient $\square$ $\qquad \frac{40}{10} = \square$ $\qquad 4\overline{)40}$ quotient $\square$

Activities

There are forty blocks.

There are ten blocks in each group.

How many groups of blocks are there? ____

Use the numbers below to make equations.

4 10 40

$\bigcirc \div 10 = \bigcirc$

$40 \div \bigcirc = \bigcirc$

$\bigcirc \div \bigcirc = 4$

$\bigcirc \div 4 = \bigcirc$

Make equations with the numbers below.

4 10 40

$\square \overline{)\ \square\ }$ with $\square$ above; $\dfrac{\square}{\square} = \square$

$\square \overline{)\ \square\ }$ with $\square$ above; $\dfrac{\square}{\square} = \square$

Divide the balloons into ten even groups by circling each group.

Write the equation to match the picture. ______ ÷ ______ = ______

Fill in the Answer

Fill in the blanks to complete the equations.

$44 \div 4 = \square$ $\quad$ $44 \div \square = 11$ $\quad$ $44 \div 4 = \square$

$44 \div \square = 4$ $\quad$ $\square \div 4 = 11$ $\quad$ $44 \div \square = 11$

$\square \div 11 = 4$ $\quad$ $44 \div \square = 4$ $\quad$ $\square \div 4 = 11$

$44 \div 11 = \square$ $\quad$ $\square \div 11 = 4$ $\quad$ $44 \div 11 = \square$

$\begin{array}{r} 44 \\ \div \; \square \\ \hline 11 \end{array}$ $\quad$ $\begin{array}{r} 44 \\ \div \; 11 \\ \hline \square \end{array}$ $\quad$ $\begin{array}{r} \square \\ \div \; 11 \\ \hline 4 \end{array}$ $\quad$ $\begin{array}{r} 44 \\ \div \; 4 \\ \hline \square \end{array}$ $\quad$ $\begin{array}{r} \square \\ \div \; 4 \\ \hline 11 \end{array}$ $\quad$ $\begin{array}{r} 44 \\ \div \; \square \\ \hline 4 \end{array}$

$4\overline{)44}$ (quotient: $\square$) $\quad$ $\frac{44}{11} = \square$ $\quad$ $11\overline{)44}$ (quotient: $\square$) $\quad$ $\frac{44}{4} = \square$ $\quad$ $4\overline{)44}$ (quotient: $\square$)

Activities

There are forty-four ants.

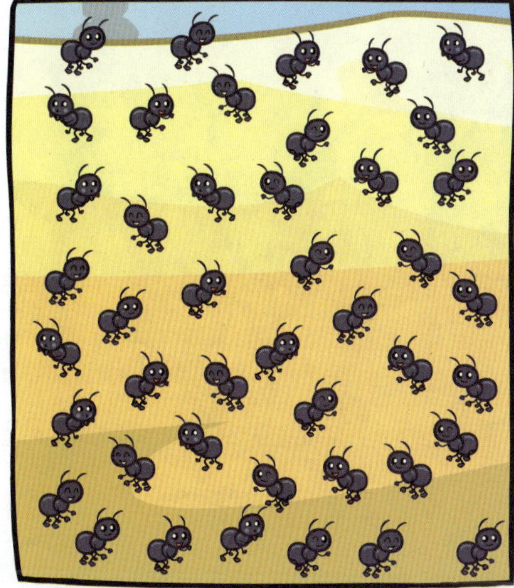

Eleven ants live in each cactus.

How many cacti do we need for each ant to have a home? ______

Use the numbers below to make equations.

4 11 44

44 ÷ ___ = ___

___ ÷ ___ = 11

___ ÷ ___ = 4

___ ÷ 11 = ___

Make equations with the numbers below.

4 11 44

Divide the cacti into eleven even groups by circling each group.

Write the equation to match the picture. ______ ÷ ______ = ______

Fill in the Answer

Fill in the blanks to complete the equations.

$48 \div 4 = \square$

$\square \div 12 = 4$

$\square \div 12 = 4$

$48 \div 12 = \square$

$48 \div \square = 4$

$48 \div \square = 4$

$48 \div \square = 12$

$\square \div 4 = 12$

$48 \div 12 = \square$

$\square \div 4 = 12$

$48 \div \square = 12$

$48 \div 4 = \square$

$\square \div 12 = 4$

$48 \div 4 = \square$

$48 \div \square = 4$

$\square \div 4 = 12$

$48 \div \square = 12$

$48 \div 12 = \square$

$4\overline{)48}$ = $\square$

$\frac{48}{4} = \square$

$12\overline{)48}$ = $\square$

$\frac{48}{12} = \square$

$4\overline{)48}$ = $\square$

Activities

There are forty-eight carrots.

There are four rabbits.

How many carrots will each rabbit get if they are shared evenly? ______

Use the numbers below to make equations.

Make equations with the numbers below.

Divide the berries into twelve even groups by circling each group.

Write the equation to match the picture. ______ ÷ ______ = ______

Fill in the Answer

Fill in the blanks to complete the equations.

$\square \div 5 = 5$ $25 \div 5 = \square$ $25 \div 5 = \square$

$25 \div 5 = \square$ $\square \div 5 = 5$ $\square \div 5 = 5$

$25 \div \square = 5$ $25 \div \square = 5$ $25 \div \square = 5$

$25 \div 5 = \square$ $\square \div 5 = 5$ $25 \div \square = 5$

$\begin{array}{r} 25 \\ \div\ 5 \\ \hline \square \end{array}$ $\begin{array}{r} \square \\ \div\ 5 \\ \hline 5 \end{array}$ $\begin{array}{r} 25 \\ \div\ \square \\ \hline 5 \end{array}$ $\begin{array}{r} 25 \\ \div\ 5 \\ \hline \square \end{array}$ $\begin{array}{r} 25 \\ \div\ \square \\ \hline 5 \end{array}$ $\begin{array}{r} 25 \\ \div\ 5 \\ \hline \square \end{array}$

$5\overline{)25}$ (quotient: $\square$) $\frac{25}{5} = \square$ $5\overline{)25}$ (quotient: $\square$) $\frac{25}{5} = \square$ $5\overline{)25}$ (quotient: $\square$)

Activities

There are twenty-five cones.

There are five puddles of mud.

When divided evenly, how many cones can be put around each puddle of mud? ___

Use the numbers below to make equations.

5 5 25

◯ ÷ ◯ = 5

◯ ÷ 5 = ◯

◯ ÷ 5 = ◯

25 ÷ ◯ = ◯

Make equations with the numbers below.

5 5 25

Divide the cones into five even groups by circling each group.

Write the equation to match the picture. ______ ÷ ______ = ______

Fill in the Answer

Fill in the blanks to complete the equations.

$\square \div 5 = 6$ $\quad 30 \div 5 = \square$ $\quad \square \div 5 = 6$

$30 \div 6 = \square$ $\quad 30 \div \square = 6$ $\quad 30 \div 5 = \square$

$30 \div \square = 6$ $\quad \square \div 6 = 5$ $\quad \square \div 6 = 5$

$30 \div \square = 5$ $\quad 30 \div 6 = \square$ $\quad 30 \div \square = 5$

$\begin{array}{r} 30 \\ \div\ \square \\ \hline 6 \end{array}$ $\quad \begin{array}{r} 30 \\ \div\ 6 \\ \hline \square \end{array}$ $\quad \begin{array}{r} \square \\ \div\ 5 \\ \hline 6 \end{array}$ $\quad \begin{array}{r} 30 \\ \div\ 5 \\ \hline \square \end{array}$ $\quad \begin{array}{r} \square \\ \div\ 6 \\ \hline 5 \end{array}$ $\quad \begin{array}{r} 30 \\ \div\ \square \\ \hline 5 \end{array}$

$\begin{array}{r} \square \\ 5\overline{)30} \end{array}$ $\quad \frac{30}{6} = \square$ $\quad \begin{array}{r} \square \\ 6\overline{)30} \end{array}$ $\quad \frac{30}{5} = \square$ $\quad \begin{array}{r} \square \\ 5\overline{)30} \end{array}$

Activities

There are thirty asteroids.

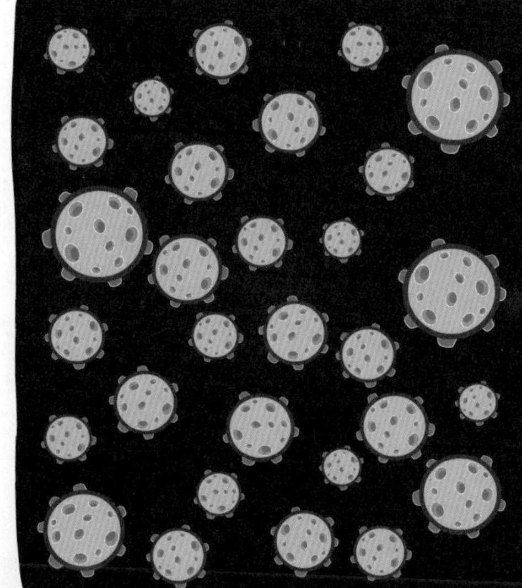

Six asteroids are in each group.

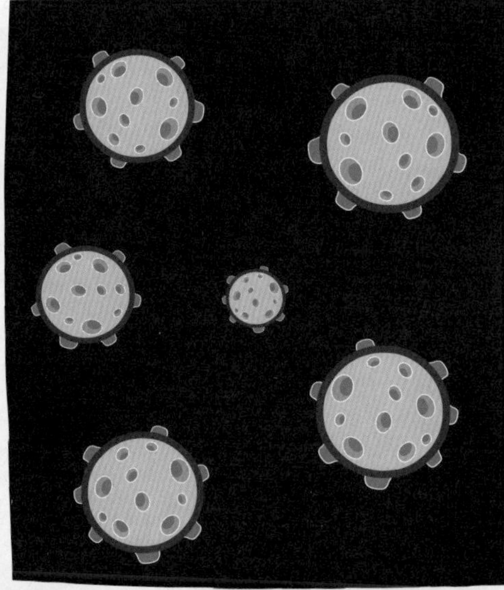

How many groups of asteroids are there in all? _____

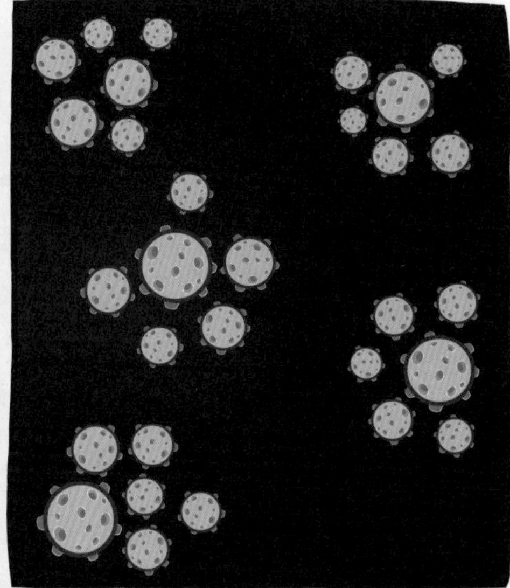

Use the numbers below to make equations.

5 6 30

30 ÷ ___ = ___

___ ÷ 6 = ___

___ ÷ 5 = ___

___ ÷ ___ = 5

Make equations with the numbers below.

5 6 30

Divide the asteroids into five even groups by circling each group.

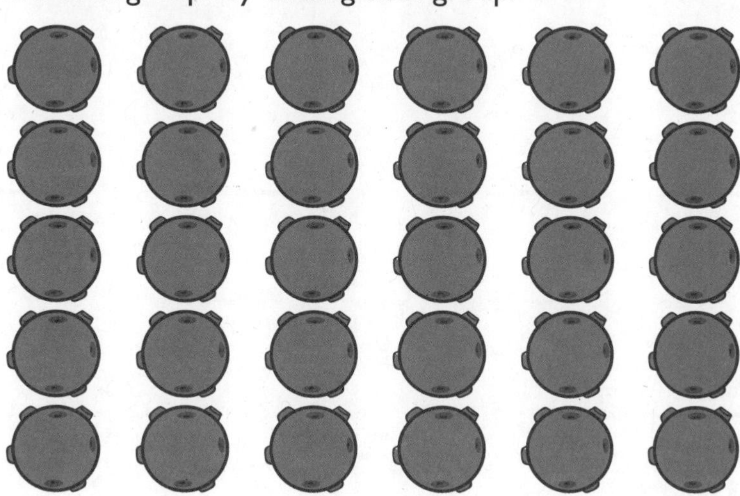

Write the equation to match the picture. _______ ÷ _______ = _______

Fill in the Answer

Fill in the blanks to complete the equations.

$35 \div \square = 5$	$35 \div 7 = \square$	$35 \div \square = 5$
$\square \div 5 = 7$	$\square \div 7 = 5$	$35 \div 7 = \square$
$35 \div 5 = \square$	$35 \div 7 = \square$	$35 \div \square = 7$
$\square \div 5 = 7$	$35 \div \square = 7$	$\square \div 5 = 7$

$$\begin{array}{r}\square\\ \div\ 5\\ \hline 7\end{array}\qquad \begin{array}{r}35\\ \div\ 7\\ \hline \square\end{array}\qquad \begin{array}{r}\square\\ \div\ 7\\ \hline 5\end{array}\qquad \begin{array}{r}\square\\ \div\ 5\\ \hline 7\end{array}\qquad \begin{array}{r}35\\ \div\ 5\\ \hline \square\end{array}\qquad \begin{array}{r}35\\ \div\ \square\\ \hline 5\end{array}$$

$$\begin{array}{r}\square\\ 5\overline{)35}\end{array}\qquad \frac{35}{7} = \square\qquad \begin{array}{r}\square\\ 7\overline{)35}\end{array}\qquad \frac{35}{5} = \square\qquad \begin{array}{r}\square\\ 5\overline{)35}\end{array}$$

Activities

There are thirty-five flowers.

We will put seven flowers in each vase.

How many vases will we need? _____

Use the numbers below to make equations.

5 7 35

☐ ÷ ☐ = 7

35 ÷ ☐ = ☐

☐ ÷ 7 = ☐

☐ ÷ ☐ = 5

Make equations with the numbers below.

5 7 35

Divide the bow ties into five even groups by circling each group.

Write the equation to match the picture. _______ ÷ _______ = _______

Fill in the Answer

Fill in the blanks to complete the equations.

$40 \div \square = 8$ $40 \div 5 = \square$ $40 \div 8 = \square$

$\square \div 8 = 5$ $40 \div \square = 8$ $40 \div \square = 5$

$40 \div \square = 8$ $\square \div 8 = 5$ $\square \div 5 = 8$

$\square \div 8 = 5$ $40 \div 5 = \square$ $40 \div 8 = \square$

$$\begin{array}{r} 40 \\ \div\ 5 \\ \hline \square \end{array} \quad \begin{array}{r} 40 \\ \div\ \square \\ \hline 8 \end{array} \quad \begin{array}{r} \square \\ \div\ 5 \\ \hline 8 \end{array} \quad \begin{array}{r} \square \\ \div\ 8 \\ \hline 5 \end{array} \quad \begin{array}{r} 40 \\ \div\ 8 \\ \hline \square \end{array} \quad \begin{array}{r} \square \\ \div\ 5 \\ \hline 8 \end{array}$$

$$5\overline{)40} \quad \frac{40}{5} = \square \quad 8\overline{)40} \quad \frac{40}{8} = \square \quad 5\overline{)40}$$

(each long division has a blank box above for the answer)

Activities

There are forty balls.

We put eight balls in each bowl.

How many bowls do we need? _____

Use the numbers below to make equations.

5 8 40

◯ ÷ ◯ = 5

40 ÷ ◯ = ◯

◯ ÷ 8 = ◯

◯ ÷ ◯ = 8

Make equations with the numbers below.

5 8 40

Divide the gum balls into five even groups by circling each group.

Write the equation to match the picture. _______ ÷ _______ = _______

Fill in the Answer

Fill in the blanks to complete the equations.

$45 \div 9 = \square$ $45 \div \square = 9$ $45 \div \square = 5$

$45 \div \square = 5$ $\square \div 5 = 9$ $45 \div 5 = \square$

$\square \div 9 = 5$ $45 \div 9 = \square$ $\square \div 5 = 9$

$45 \div 5 = \square$ $\square \div 9 = 5$ $45 \div \square = 9$

$\begin{array}{r} 45 \\ \div \;\square \\ \hline 5 \end{array}$ $\begin{array}{r} 45 \\ \div \;\; 9 \\ \hline \square \end{array}$ $\begin{array}{r} \square \\ \div \;\; 9 \\ \hline 5 \end{array}$ $\begin{array}{r} 45 \\ \div \;\; 5 \\ \hline \square \end{array}$ $\begin{array}{r} \square \\ \div \;\; 5 \\ \hline 9 \end{array}$ $\begin{array}{r} 45 \\ \div \;\square \\ \hline 9 \end{array}$

$9\overline{)45}$ (answer: $\square$) $\frac{45}{5} = \square$ $5\overline{)45}$ (answer: $\square$) $\frac{45}{9} = \square$ $5\overline{)45}$ (answer: $\square$)

Activities

There are forty-five rocks.

There are five potholes on the road.

When divided evenly, how many rocks will we put into each pothole? _____

Use the numbers below to make equations.

5 9 45

45 ÷ ☐ = ☐

☐ ÷ 5 = ☐

☐ ÷ ☐ = 9

Make equations with the numbers below.

5 9 45

Divide the helmets into nine even groups by circling each group.

Write the equation to match the picture. _______ ÷ _______ = _______

Fill in the Answer

Fill in the blanks to complete the equations.

$50 \div \square = 10$ | $\square \div 5 = 10$ | $50 \div 10 = \square$

$50 \div 10 = \square$ | $50 \div \square = 5$ | $50 \div \square = 10$

$50 \div \square = 5$ | $\square \div 10 = 5$ | $\square \div 5 = 10$

$\square \div 10 = 5$ | $50 \div 5 = \square$ | $50 \div 5 = \square$

$$\begin{array}{r} 50 \\ \div\ \square \\ \hline 10 \end{array} \quad \begin{array}{r} 50 \\ \div\ \square \\ \hline 5 \end{array} \quad \begin{array}{r} 50 \\ \div\ 5 \\ \hline \square \end{array} \quad \begin{array}{r} \square \\ \div\ 10 \\ \hline 5 \end{array} \quad \begin{array}{r} \square \\ \div\ 5 \\ \hline 10 \end{array} \quad \begin{array}{r} 50 \\ \div\ 10 \\ \hline \square \end{array}$$

$$5 \overline{)\,50} = \square \quad \frac{50}{5} = \square \quad 10 \overline{)\,50} = \square \quad \frac{50}{10} = \square \quad 5 \overline{)\,50} = \square$$

Activities

There are fifty rocks. | There are ten rocks in each group. | How many groups of rocks are there? ___

Use the numbers below to make equations.

5 10 50

___	÷	5	=	___
50	÷	___	=	___
___	÷	10	=	___
___	÷	___	=	10

Make equations with the numbers below.

5 10 50

Divide the trees into ten even groups by circling each group.

Write the equation to match the picture. ______ ÷ ______ = ______

Fill in the Answer

Fill in the blanks to complete the equations.

$\square \div 11 = 5$ $\qquad 55 \div \square = 5$ $\qquad 55 \div 5 = \square$

$55 \div \square = 11$ $\qquad \square \div 11 = 5$ $\qquad 55 \div \square = 5$

$55 \div 5 = \square$ $\qquad 55 \div 11 = \square$ $\qquad \square \div 11 = 5$

$\square \div 5 = 11$ $\qquad 55 \div \square = 11$ $\qquad 55 \div 11 = \square$

$$\begin{array}{r}55\\ \div\ \square\\ \hline 11\end{array}\qquad \begin{array}{r}55\\ \div\ 5\\ \hline \square\end{array}\qquad \begin{array}{r}\square\\ \div\ 11\\ \hline 5\end{array}\qquad \begin{array}{r}55\\ \div\ 11\\ \hline \square\end{array}\qquad \begin{array}{r}\square\\ \div\ 5\\ \hline 11\end{array}\qquad \begin{array}{r}\square\\ \div\ 5\\ \hline 11\end{array}$$

$$\begin{array}{r}\square\\ 5\overline{)55}\end{array}\qquad \frac{55}{5} = \square\qquad \begin{array}{r}\square\\ 11\overline{)55}\end{array}\qquad \frac{55}{11} = \square\qquad \begin{array}{r}\square\\ 5\overline{)55}\end{array}$$

Activities

There are fifty-five flowers.

There are five pieces of string.

When divided evenly, how many flowers can we put on each lei? ______

Use the numbers below to make equations.

5 11 55

◯ ÷ ◯ = 11

55 ÷ ◯ = ◯

◯ ÷ 11 = ◯

◯ ÷ ◯ = 5

Make equations with the numbers below.

5 11 55

☐) ☐ with ☐ on top ☐ / ☐ = ☐

☐) ☐ with ☐ on top ☐ / ☐ = ☐

Divide the coconuts into eleven even groups by circling each group.

Write the equation to match the picture. ______ ÷ ______ = ______

Fill in the Answer

Fill in the blanks to complete the equations.

$60 \div \square = 12$ $60 \div 5 = \square$ $\square \div 5 = 12$

$\square \div 5 = 12$ $60 \div \square = 12$ $60 \div 5 = \square$

$60 \div 12 = \square$ $\square \div 5 = 12$ $\square \div 12 = 5$

$60 \div \square = 5$ $60 \div 12 = \square$ $60 \div \square = 5$

$\begin{array}{r} 60 \\ \div\ 12 \\ \hline \square \end{array}$ $\begin{array}{r} \square \\ \div\ 5 \\ \hline 12 \end{array}$ $\begin{array}{r} \square \\ \div\ 12 \\ \hline 5 \end{array}$ $\begin{array}{r} \square \\ \div\ 5 \\ \hline 12 \end{array}$ $\begin{array}{r} 60 \\ \div\ \square \\ \hline 12 \end{array}$ $\begin{array}{r} 60 \\ \div\ 5 \\ \hline \square \end{array}$

$5\overline{)60}$ (answer: $\square$) $\frac{60}{5} = \square$ $12\overline{)60}$ (answer: $\square$) $\frac{60}{12} = \square$ $5\overline{)60}$ (answer: $\square$)

Activities

There are sixty cars.

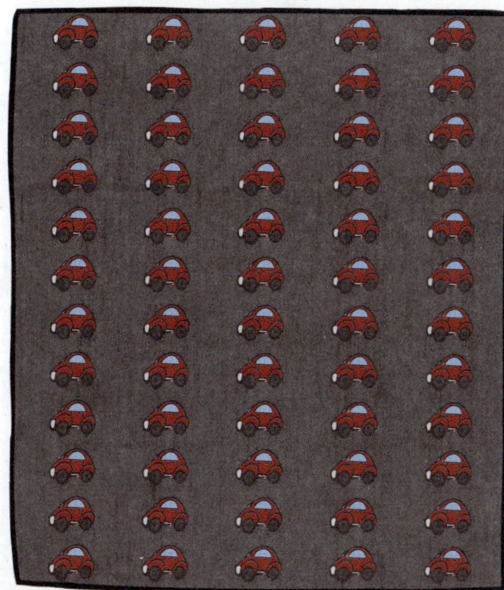

A tow truck carries five cars.

How many tow trucks will we need to carry all sixty cars? ______

Use the numbers below to make equations.

5 12 60

60 ÷ ◯ = ◯

◯ ÷ 5 = ◯

◯ ÷ 12 = ◯

◯ ÷ ◯ = 12

Make equations with the numbers below.

5 12 60

☐)☐☐ ☐/☐ = ☐

☐)☐☐ ☐/☐ = ☐

Divide the cars into twelve even groups by circling each group.

Write the equation to match the picture. ______ ÷ ______ = ______

Fill in the Answer

Fill in the blanks to complete the equations.

$36 \div 6 = \square$	$\square \div 6 = 6$	$36 \div \square = 6$
$36 \div \square = 6$	$36 \div 6 = \square$	$36 \div 6 = \square$
$\square \div 6 = 6$	$36 \div \square = 6$	$\square \div 6 = 6$
$36 \div \square = 6$	$\square \div 6 = 6$	$36 \div 6 = \square$

$$\begin{array}{r} 36 \\ \div\ \square \\ \hline 6 \end{array} \qquad \begin{array}{r} \square \\ \div\ 6 \\ \hline 6 \end{array} \qquad \begin{array}{r} 36 \\ \div\ 6 \\ \hline \square \end{array} \qquad \begin{array}{r} 36 \\ \div\ 6 \\ \hline \square \end{array} \qquad \begin{array}{r} 36 \\ \div\ \square \\ \hline 6 \end{array} \qquad \begin{array}{r} \square \\ \div\ 6 \\ \hline 6 \end{array}$$

$$\begin{array}{r} \square \\ 6\overline{)36} \end{array} \qquad \frac{36}{6} = \square \qquad \begin{array}{r} \square \\ 6\overline{)36} \end{array} \qquad \frac{36}{6} = \square \qquad \begin{array}{r} \square \\ 6\overline{)36} \end{array}$$

Activities

There are thirty-six spokes.

There are six tires

Each tire has the same number of spokes. How many spokes are on each tire? ______

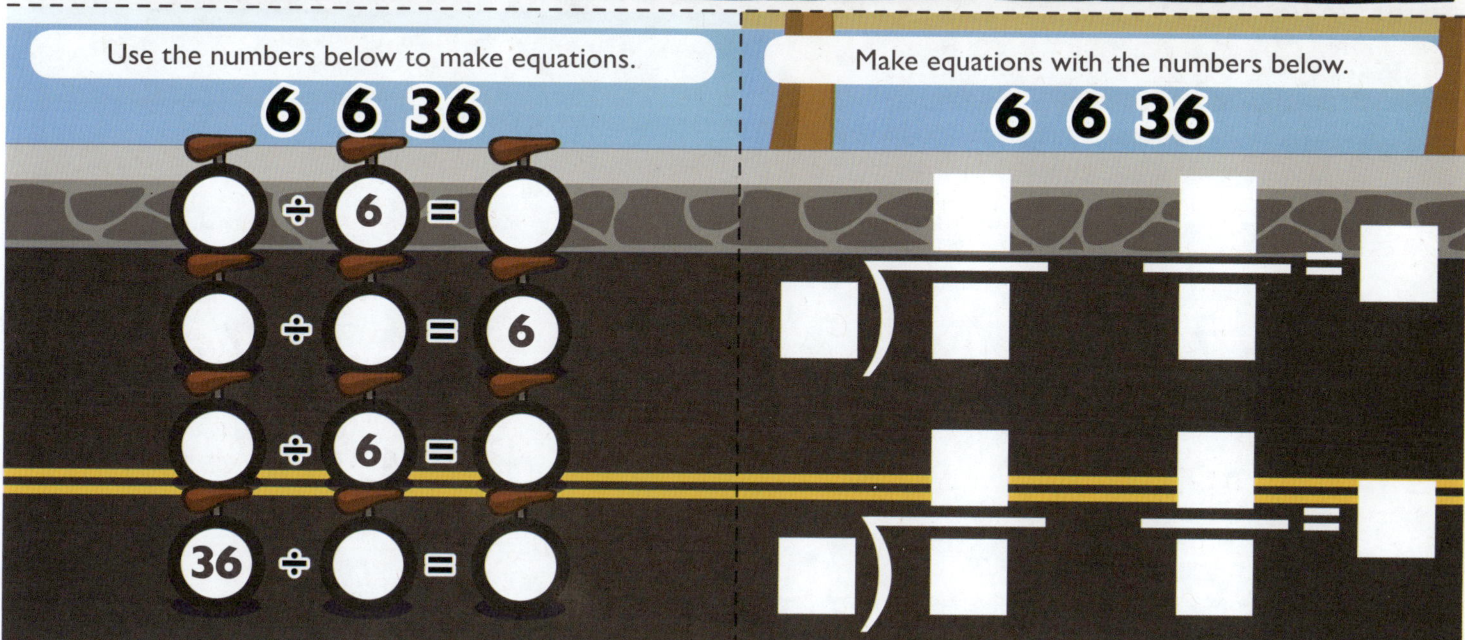

Divide the wheels into six even groups by circling each group.

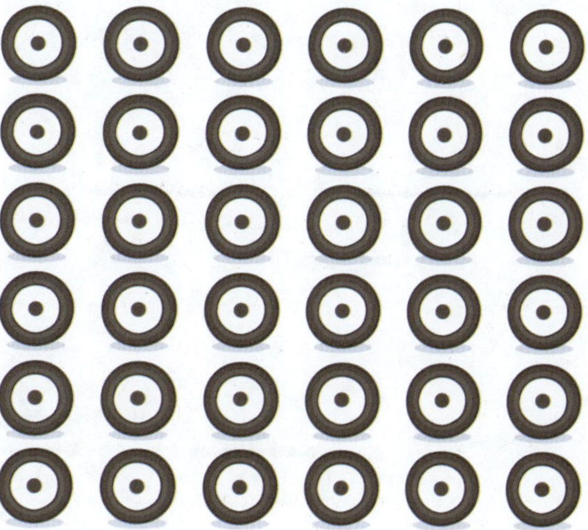

Write the equation to match the picture. ______ ÷ ______ = ______

Fill in the Answer

Fill in the blanks to complete the equations.

$42 \div \square = 6$ $42 \div 7 = \square$ $42 \div 6 = \square$

$42 \div 7 = \square$ $\square \div 6 = 7$ $\square \div 7 = 6$

$\square \div 6 = 7$ $42 \div \square = 6$ $42 \div \square = 7$

$42 \div \square = 6$ $42 \div 7 = \square$ $\square \div 6 = 7$

$\begin{array}{r} 42 \\ \div\ 7 \\ \hline \square \end{array}$ $\begin{array}{r} \square \\ \div\ 7 \\ \hline 6 \end{array}$ $\begin{array}{r} 42 \\ \div\ \square \\ \hline 12 \end{array}$ $\begin{array}{r} \square \\ \div\ 6 \\ \hline 7 \end{array}$ $\begin{array}{r} 42 \\ \div\ 6 \\ \hline \square \end{array}$ $\begin{array}{r} \square \\ \div\ 6 \\ \hline 7 \end{array}$

$6\overline{)42}$ (answer: $\square$) $\frac{42}{7} = \square$ $7\overline{)42}$ (answer: $\square$) $\frac{42}{6} = \square$ $6\overline{)42}$ (answer: $\square$)

Activities

There are forty-two nails.

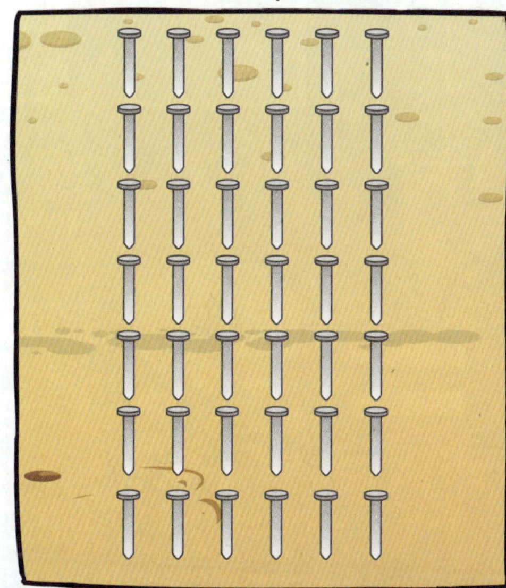

There are six buckets.

How many nails will be in each bucket when divided evenly? ______

Use the numbers below to make equations.

6 7 42

___ ÷ ___ = 7

42 ÷ ___ = ___

___ ÷ 7 = ___

___ ÷ ___ = 6

Make equations with the numbers below.

6 7 42

Divide the hammers into seven even groups by circling each group.

Write the equation to match the picture. ______ ÷ ______ = ______

Fill in the Answer

Fill in the blanks to complete the equations.

$48 \div 6 = \square$ $\quad$ $48 \div 8 = \square$ $\quad$ $\square \div 6 = 8$

$48 \div \square = 6$ $\quad$ $\square \div 8 = 6$ $\quad$ $48 \div 8 = \square$

$\square \div 6 = 8$ $\quad$ $48 \div 6 = \square$ $\quad$ $\square \div 8 = 6$

$48 \div \square = 8$ $\quad$ $48 \div \square = 6$ $\quad$ $48 \div \square = 8$

$\begin{array}{r} 48 \\ \div\ 6 \\ \hline \square \end{array}$ $\quad$ $\begin{array}{r} \square \\ \div\ 8 \\ \hline 6 \end{array}$ $\quad$ $\begin{array}{r} 48 \\ \div\ \square \\ \hline 6 \end{array}$ $\quad$ $\begin{array}{r} 48 \\ \div\ 8 \\ \hline \square \end{array}$ $\quad$ $\begin{array}{r} 48 \\ \div\ \square \\ \hline 8 \end{array}$ $\quad$ $\begin{array}{r} 48 \\ \div\ 6 \\ \hline \square \end{array}$

$\begin{array}{r} \square \\ 6\overline{)48} \end{array}$ $\quad$ $\dfrac{48}{6} = \square$ $\quad$ $\begin{array}{r} \square \\ 8\overline{)48} \end{array}$ $\quad$ $\dfrac{48}{8} = \square$ $\quad$ $\begin{array}{r} \square \\ 6\overline{)48} \end{array}$

Activities

There are forty-eight darts.

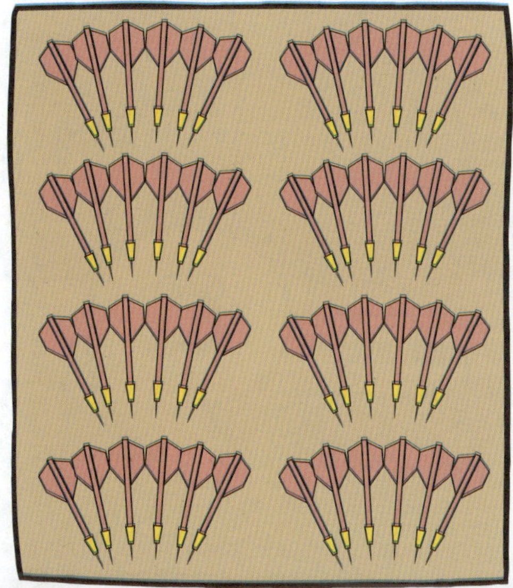

There are six dartboards.

Each board gets the same number of darts, How many darts does each board get? ____

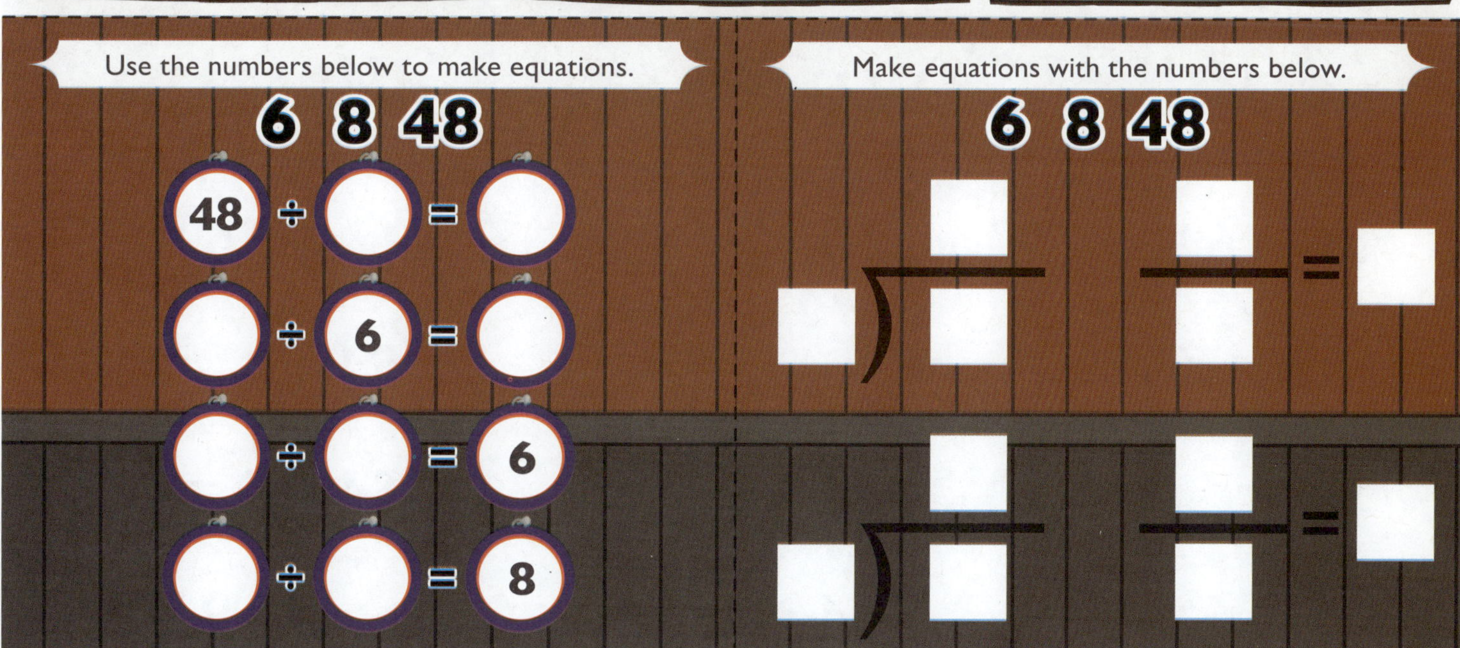

Divide the dartboards into eight even groups by circling each group.

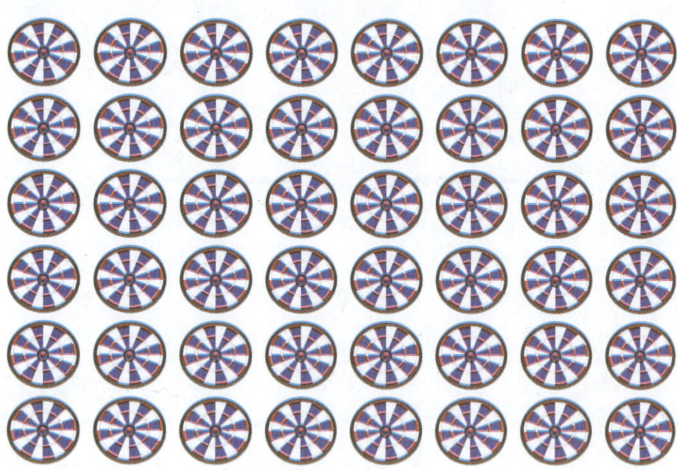

Write the equation to match the picture. ______ ÷ ______ = ______

Fill in the Answer

Fill in the blanks to complete the equations.

$54 \div 6 = \square$ $\qquad 54 \div \square = 9$ $\qquad \square \div 6 = 9$

$\square \div 6 = 9$ $\qquad \square \div 9 = 6$ $\qquad 54 \div \square = 9$

$54 \div \square = 6$ $\qquad 54 \div 6 = \square$ $\qquad 54 \div 9 = \square$

$54 \div 6 = \square$ $\qquad 54 \div \square = 6$ $\qquad \square \div 9 = 6$

$$\begin{array}{r} 54 \\ \div\ \square \\ \hline 9 \end{array} \qquad \begin{array}{r} 54 \\ \div\ 6 \\ \hline \square \end{array} \qquad \begin{array}{r} \square \\ \div\ 6 \\ \hline 9 \end{array} \qquad \begin{array}{r} 54 \\ \div\ 9 \\ \hline \square \end{array} \qquad \begin{array}{r} \square \\ \div\ 9 \\ \hline 6 \end{array} \qquad \begin{array}{r} 54 \\ \div\ \square \\ \hline 6 \end{array}$$

$$\begin{array}{r} \square \\ 6\overline{)54} \end{array} \qquad \frac{54}{6} = \square \qquad \begin{array}{r} \square \\ 9\overline{)54} \end{array} \qquad \frac{54}{9} = \square \qquad \begin{array}{r} \square \\ 9\overline{)54} \end{array}$$

Activities

There are fifty-four rocks.

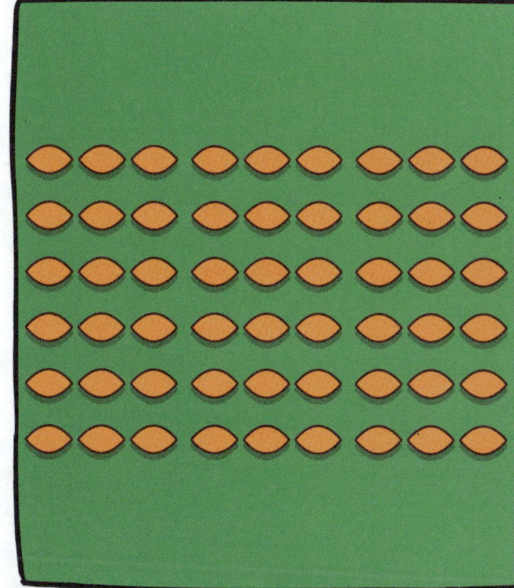

There are six holes.

How many rocks will be in each hole when divided evenly? ______

Use the numbers below to make equations.

6 9 54

☐ ÷ 9 = ☐

☐ ÷ ☐ = 6

☐ ÷ 6 = ☐

54 ÷ ☐ = ☐

Make equations with the numbers below.

6 9 54

Divide the water containers into nine even groups by circling each group.

Write the equation to match the picture. ______ ÷ ______ = ______

Fill in the Answer

Fill in the blanks to complete the equations.

$60 \div \square = 10$ $\square \div 10 = 6$ $\square \div 6 = 10$

$60 \div 6 = \square$ $60 \div 10 = \square$ $60 \div 10 = \square$

$\square \div 6 = 10$ $60 \div \square = 10$ $\square \div 10 = 6$

$60 \div \square = 6$ $60 \div 6 = \square$ $60 \div \square = 6$

$\begin{array}{r} \square \\ \div\ 6 \\ \hline 10 \end{array}$ $\begin{array}{r} 60 \\ \div\ 10 \\ \hline \square \end{array}$ $\begin{array}{r} 60 \\ \div\ \square \\ \hline 10 \end{array}$ $\begin{array}{r} \square \\ \div\ 10 \\ \hline 6 \end{array}$ $\begin{array}{r} 60 \\ \div\ 6 \\ \hline \square \end{array}$ $\begin{array}{r} 60 \\ \div\ \square \\ \hline 6 \end{array}$

$6\overline{)60}$ with $\square$ above $\frac{60}{10} = \square$ $10\overline{)60}$ with $\square$ above $\frac{60}{6} = \square$ $6\overline{)60}$ with $\square$ above

Activities

There are sixty sandwiches.

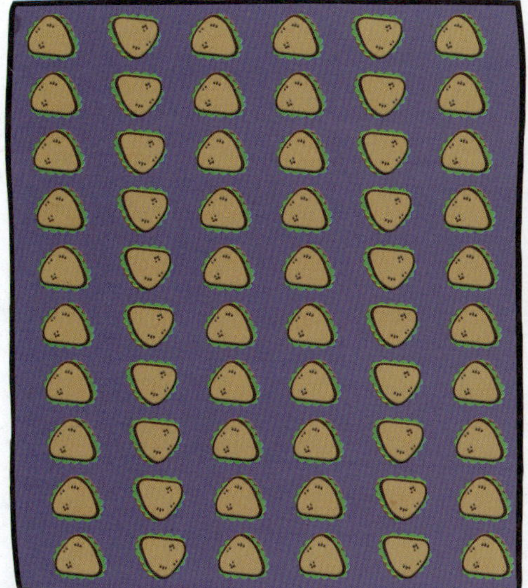

There are six lunch boxes.

When divided equally, how many sandwiches will be in each box? _____

Use the numbers below to make equations.

6 10 60

☐ ÷ 10 = ☐

☐ ÷ ☐ = 10

☐ ÷ 6 = ☐

60 ÷ ☐ = ☐

Make equations with the numbers below.

6 10 60

Divide the lunch boxes into ten even groups by circling each group.

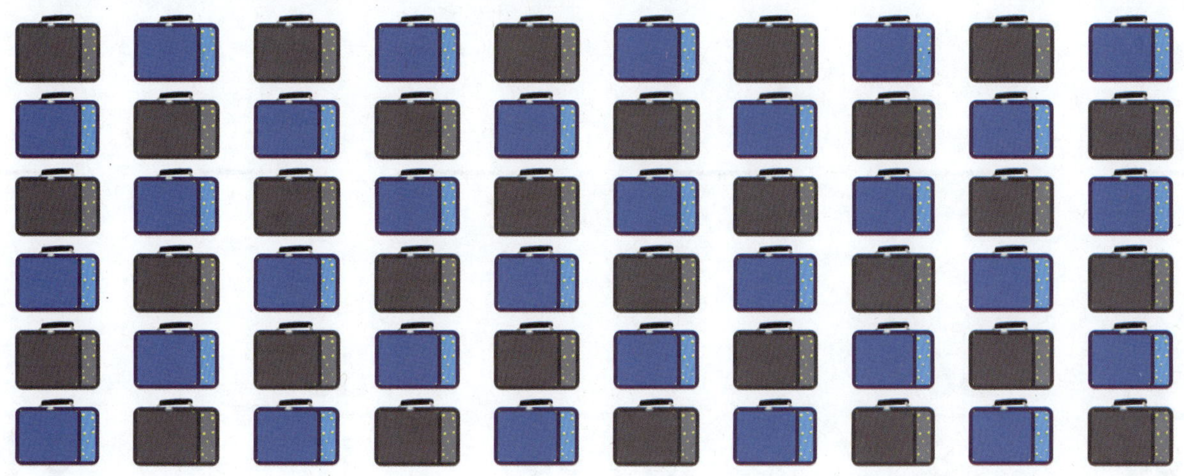

Write the equation to match the picture. ______ ÷ ______ = ______

Fill in the Answer

Fill in the blanks to complete the equations.

___ ÷ 6 = 11	66 ÷ 11 = ___	___ ÷ 6 = 11
66 ÷ 6 = ___	66 ÷ ___ = 6	66 ÷ 11 = ___
66 ÷ ___ = 6	___ ÷ 11 = 6	___ ÷ 11 = 6
66 ÷ ___ = 11	66 ÷ 6 = ___	66 ÷ ___ = 11

66	66	___	66	66	66
÷ ___	÷ 11	÷ 6	÷ 6	÷ ___	÷ ___
11	___	11	___	6	11

___		___		___
$6\overline{)66}$	$\frac{66}{6}$ = ___	$11\overline{)66}$	$\frac{66}{11}$ = ___	$6\overline{)66}$

Activities

There are sixty-six keys.

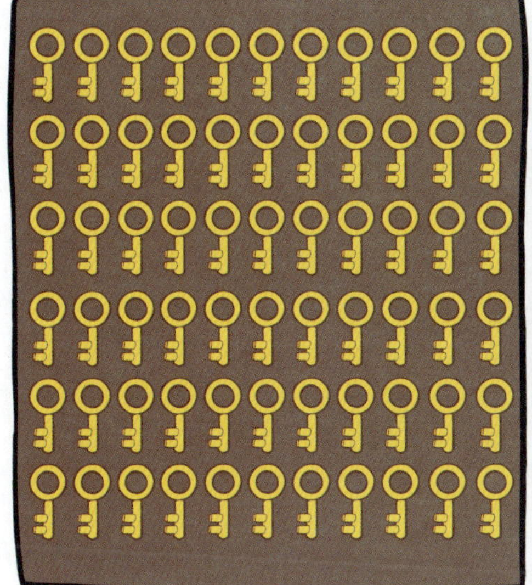

There are six key rings.

How many keys will fit on each key ring when divided evenly? _____

Use the numbers below to make equations.

6 11 66

Make equations with the numbers below.

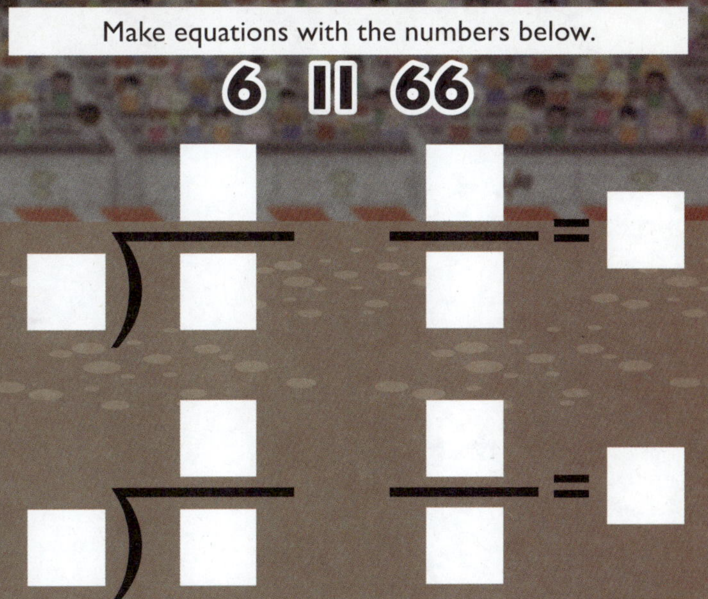

Divide the monster trucks into eleven even groups by circling each group.

Write the equation to match the picture. _____ ÷ _____ = _____

Fill in the Answer

Fill in the blanks to complete the equations.

$72 \div 6 = \square$ $72 \div \square = 6$ $72 \div 6 = \square$

$72 \div \square = 12$ $\square \div 12 = 6$ $72 \div \square = 6$

$\square \div 6 = 12$ $72 \div \square = 12$ $\square \div 12 = 6$

$72 \div 12 = \square$ $\square \div 6 = 12$ $72 \div 12 = \square$

$\begin{array}{r} 72 \\ \div\ \square \\ \hline 12 \end{array}$ $\begin{array}{r} 72 \\ \div\ 12 \\ \hline \square \end{array}$ $\begin{array}{r} \square \\ \div\ 12 \\ \hline 6 \end{array}$ $\begin{array}{r} 72 \\ \div\ 6 \\ \hline \square \end{array}$ $\begin{array}{r} \square \\ \div\ 6 \\ \hline 12 \end{array}$ $\begin{array}{r} 72 \\ \div\ \square \\ \hline 6 \end{array}$

$6\overline{)72}$ (answer: $\square$) $\frac{72}{6} = \square$ $12\overline{)72}$ (answer: $\square$) $\frac{72}{12} = \square$ $6\overline{)72}$ (answer: $\square$)

Activities

There are seventy-two rocks. We put them into groups of twelve. How many groups of rocks do we have? ____

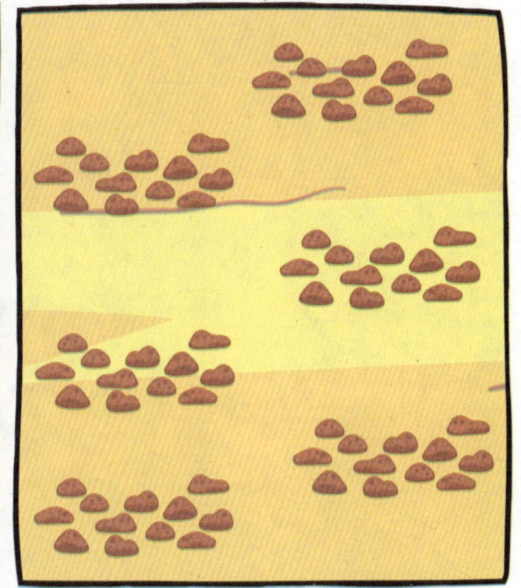

Use the numbers below to make equations.

6 12 72

☐	÷	☐	=	12
72	÷	☐	=	☐
☐	÷	6	=	☐
☐	÷	☐	=	6

Make equations with the numbers below.

6 12 72

Divide the tents into six even groups by circling each group.

Write the equation to match the picture. ______ ÷ ______ = ______

Fill in the Answer

Fill in the blanks to complete the equations.

$49 \div 7 = \square$	$\square \div 7 = 7$	$\square \div 7 = 7$
$49 \div 7 = \square$	$49 \div \square = 7$	$49 \div \square = 7$
$49 \div \square = 7$	$\square \div 7 = 7$	$49 \div 7 = \square$
$\square \div 7 = 7$	$49 \div \square = 7$	$49 \div 7 = \square$

$$\begin{array}{r}\square\\ \div\ 7\\ \hline 7\end{array}\quad \begin{array}{r}49\\ \div\ 7\\ \hline \square\end{array}\quad \begin{array}{r}49\\ \div\ \square\\ \hline 7\end{array}\quad \begin{array}{r}\square\\ \div\ 7\\ \hline 7\end{array}\quad \begin{array}{r}49\\ \div\ \square\\ \hline 7\end{array}\quad \begin{array}{r}49\\ \div\ 7\\ \hline \square\end{array}$$

$$\begin{array}{r}\square\\ 7\overline{)49}\end{array}\qquad \frac{49}{7} = \square \qquad \begin{array}{r}\square\\ 7\overline{)49}\end{array}\qquad \frac{49}{7} = \square \qquad \begin{array}{r}\square\\ 7\overline{)49}\end{array}$$

Activities

There are forty-nine rocks.

There are seven slingshots.

How many rocks are there for each slingshot get when divided equally? _____

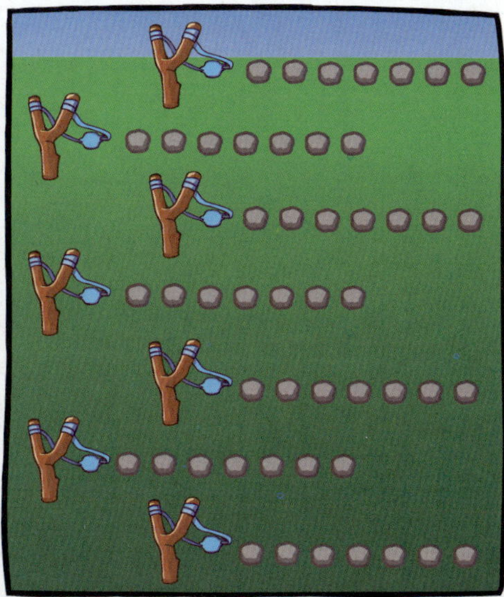

Use the numbers below to make equations.

7 7 49

$\square \div 7 = \square$

$\square \div \square = 7$

$\square \div 7 = \square$

$49 \div \square = \square$

Make equations with the numbers below.

7 7 49

$\square \overline{)\square}$ with $\square$ above $\dfrac{\square}{\square} = \square$

$\square \overline{)\square}$ with $\square$ above $\dfrac{\square}{\square} = \square$

Divide the seed bags into seven even groups by circling each group.

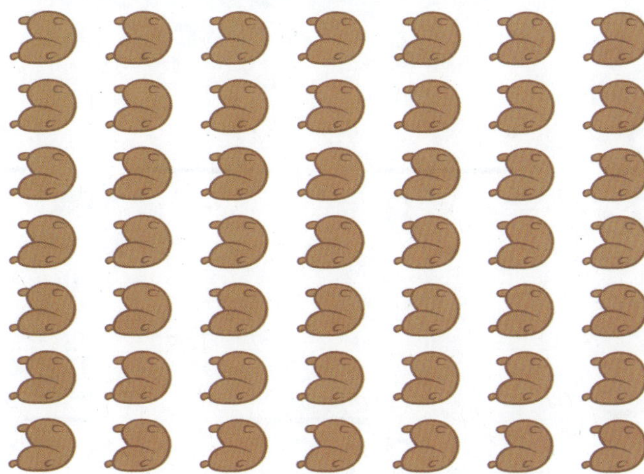

Write the equation to match the picture. ______ ÷ ______ = ______

Fill in the Answer

Fill in the blanks to complete the equations.

$\square \div 8 = 7$ $\quad$ $56 \div 8 = \square$ $\quad$ $56 \div 7 = \square$

$56 \div 8 = \square$ $\quad$ $\square \div 7 = 8$ $\quad$ $\square \div 8 = 7$

$56 \div \square = 8$ $\quad$ $56 \div \square = 7$ $\quad$ $56 \div \square = 8$

$56 \div 7 = \square$ $\quad$ $\square \div 8 = 7$ $\quad$ $56 \div \square = 7$

$$\begin{array}{r} 56 \\ \div\ \ 8 \\ \hline \square \end{array} \qquad \begin{array}{r} \square \\ \div\ \ 8 \\ \hline 7 \end{array} \qquad \begin{array}{r} 56 \\ \div\ \square \\ \hline 8 \end{array} \qquad \begin{array}{r} 56 \\ \div\ \ 7 \\ \hline \square \end{array} \qquad \begin{array}{r} 56 \\ \div\ \square \\ \hline 7 \end{array} \qquad \begin{array}{r} \square \\ \div\ \ 7 \\ \hline 8 \end{array}$$

$$8\overset{\square}{\overline{)56}} \qquad \frac{56}{8} = \square \qquad 7\overset{\square}{\overline{)56}} \qquad \frac{56}{7} = \square \qquad 8\overset{\square}{\overline{)56}}$$

Activities

There are fifty-six pins.

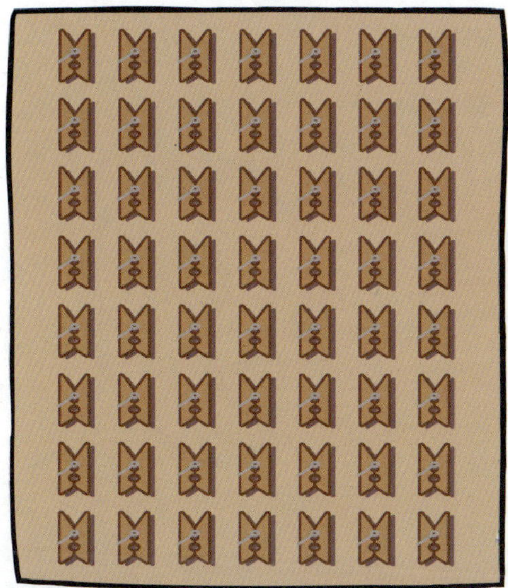

There are seven clotheslines.

When divided evenly, how many pins will we put on each clothesline? _____

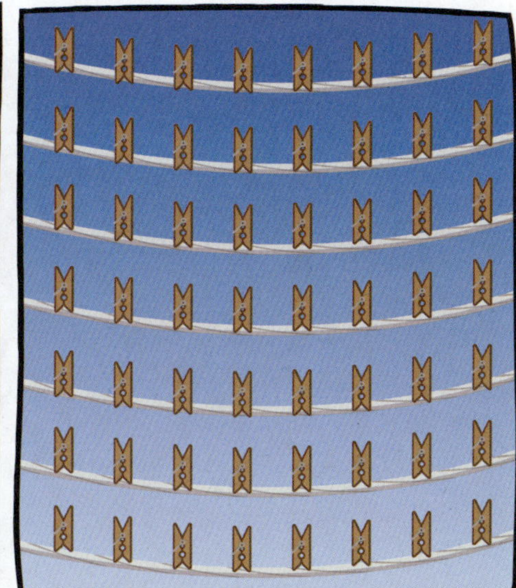

Use the numbers below to make equations.

7 8 56

56 ÷ ☐ = ☐

☐ ÷ ☐ = 8

☐ ÷ ☐ = 7

☐ ÷ 7 = ☐

Make equations with the numbers below.

7 8 56

Divide the pins into eight even groups by circling each group.

Write the equation to match the picture. _____ ÷ _____ = _____

Fill in the Answer

Fill in the blanks to complete the equations.

___ ÷ 9 = 7	63 ÷ 7 = ___	___ ÷ 9 = 7
63 ÷ 9 = ___	63 ÷ ___ = 7	63 ÷ 7 = ___
63 ÷ ___ = 7	___ ÷ 7 = 9	___ ÷ 7 = 9
63 ÷ ___ = 9	63 ÷ 9 = ___	63 ÷ ___ = 9

$\begin{array}{r} 63 \\ \div\ \square \\ \hline 9 \end{array}$ $\begin{array}{r} 63 \\ \div\ 9 \\ \hline \square \end{array}$ $\begin{array}{r} \square \\ \div\ 9 \\ \hline 7 \end{array}$ $\begin{array}{r} 63 \\ \div\ 7 \\ \hline \square \end{array}$ $\begin{array}{r} 63 \\ \div\ \square \\ \hline 7 \end{array}$ $\begin{array}{r} 63 \\ \div\ \square \\ \hline 9 \end{array}$

$7\overline{)63}$ (answer: $\square$) $\frac{63}{9} = \square$ $9\overline{)63}$ (answer: $\square$) $\frac{63}{7} = \square$ $7\overline{)63}$ (answer: $\square$)

Activities

There are sixty-three pellets of food.

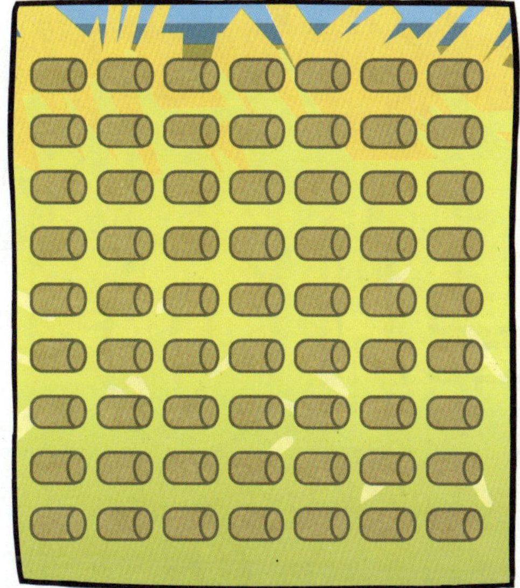

There are seven containers.

How many pellets will we put in each container when divided equally? _____

Use the numbers below to make equations.

7 9 63

___ ÷ ___ = 9

63 ÷ ___ = ___

___ ÷ 9 = ___

___ ÷ ___ = 7

Divide the hamsters into nine even groups by circling each group.

Write the equation to match the picture. _______ ÷ _______ = _______

Fill in the Answer

Fill in the blanks to complete the equations.

$70 \div \square = 10$	$70 \div 7 = \square$	$70 \div \square = 10$
$\square \div 10 = 7$	$\square \div 7 = 10$	$70 \div 7 = \square$
$70 \div 10 = \square$	$70 \div 10 = \square$	$70 \div \square = 7$
$\square \div 7 = 10$	$70 \div \square = 7$	$\square \div 10 = 7$

$$\begin{array}{r}\square\\ \div\ 10\\ \hline 7\end{array} \quad \begin{array}{r}70\\ \div\ 10\\ \hline \square\end{array} \quad \begin{array}{r}\square\\ \div\ 7\\ \hline 10\end{array} \quad \begin{array}{r}\square\\ \div\ 10\\ \hline 7\end{array} \quad \begin{array}{r}70\\ \div\ 7\\ \hline \square\end{array} \quad \begin{array}{r}70\\ \div\ \square\\ \hline 10\end{array}$$

$$7\overline{)70}^{\,\square} \quad \frac{70}{10} = \square \quad 10\overline{)70}^{\,\square} \quad \frac{70}{7} = \square \quad 7\overline{)70}^{\,\square}$$

Activities

There are seventy paint brushes.

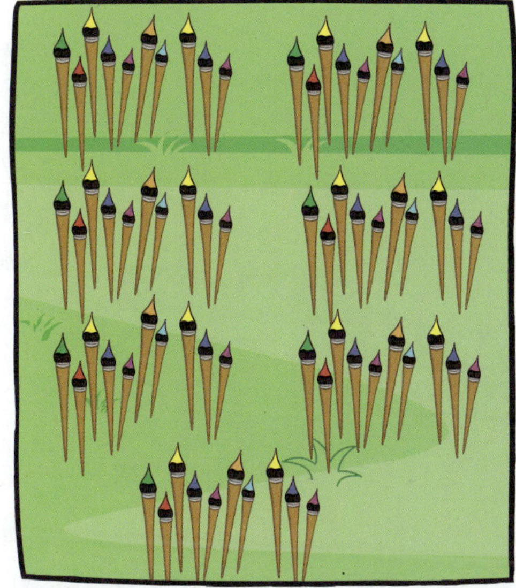

There are ten cans.

When divided evenly, how many paintbrushes will we put in each can? ___

Use the numbers below to make equations.

7 10 70

$\square \div 7 = \square$

$70 \div \square = \square$

$\square \div \square = 10$

$\square \div 10 = \square$

Make equations with the numbers below.

7 10 70

Divide the paint brushes into seven even groups by circling each group.

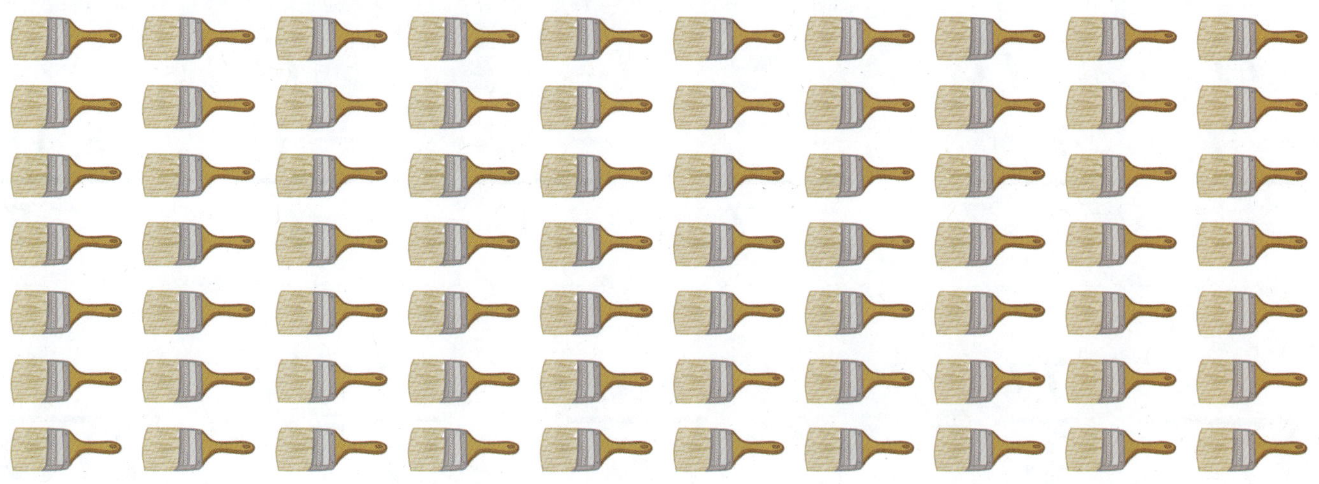

Write the equation to match the picture. ______ ÷ ______ = ______

Fill in the Answer

Fill in the blanks to complete the equations.

$77 \div \square = 7$ $77 \div 11 = \square$ $77 \div 11 = \square$

$\square \div 7 = 11$ $77 \div \square = 11$ $77 \div \square = 7$

$77 \div \square = 11$ $\square \div 7 = 11$ $\square \div 11 = 7$

$\square \div 11 = 7$ $77 \div 7 = \square$ $77 \div 11 = \square$

$\begin{array}{r} 77 \\ \div\ 7 \\ \hline \square \end{array}$ $\begin{array}{r} 77 \\ \div\ \square \\ \hline 7 \end{array}$ $\begin{array}{r} \square \\ \div\ 11 \\ \hline 7 \end{array}$ $\begin{array}{r} \square \\ \div\ 7 \\ \hline 11 \end{array}$ $\begin{array}{r} 77 \\ \div\ 11 \\ \hline \square \end{array}$ $\begin{array}{r} \square \\ \div\ 11 \\ \hline 7 \end{array}$

$7\overline{)77}$ (quotient: $\square$) $\frac{77}{11} = \square$ $11\overline{)77}$ (quotient: $\square$) $\frac{77}{7} = \square$ $7\overline{)77}$ (quotient: $\square$)

Activities

There are seventy-seven horseshoes.

There are seven hooks.

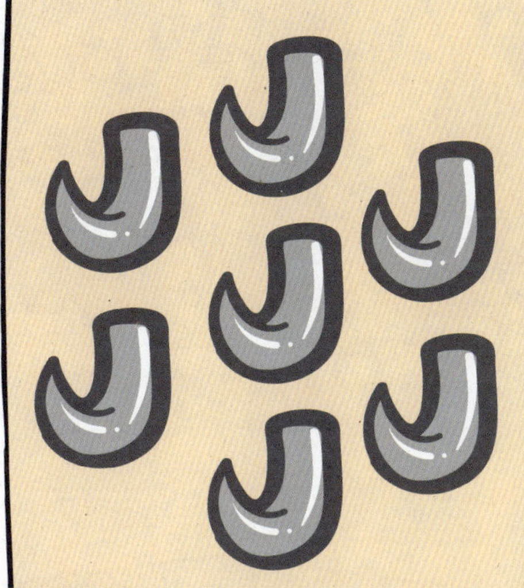

How many horseshoes do we hang from each hook when divided evenly? _____

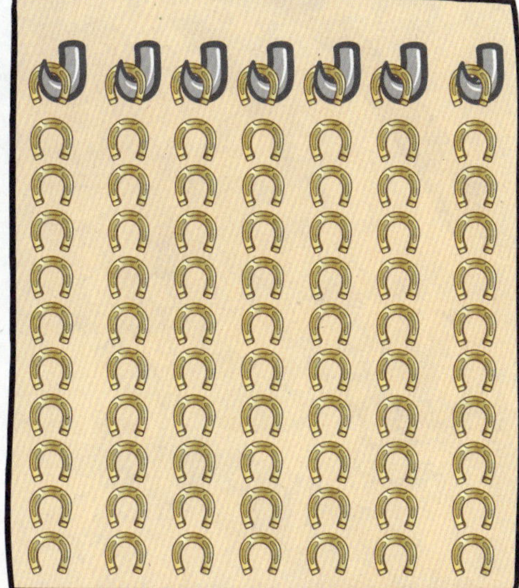

Use the numbers below to make equations.

7 11 77

___ ÷ ___ = 7

77 ÷ ___ = ___

___ ÷ 11 = ___

___ ÷ ___ = 11

Divide the hats into eleven even groups by circling each group.

Write the equation to match the picture. _______ ÷ _______ = _______

Fill in the Answer

Fill in the blanks to complete the equations.

$84 \div 12 = \square$ $\qquad 84 \div \square = 7$ $\qquad 84 \div \square = 12$

$84 \div \square = 7$ $\qquad \square \div 12 = 7$ $\qquad 84 \div 7 = \square$

$\square \div 7 = 12$ $\qquad 84 \div 12 = \square$ $\qquad \square \div 12 = 7$

$84 \div 7 = \square$ $\qquad \square \div 7 = 12$ $\qquad 84 \div \square = 12$

$\begin{array}{r} 84 \\ \div\ \square \\ \hline 7 \end{array}$ $\quad \begin{array}{r} 84 \\ \div\ 7 \\ \hline \square \end{array}$ $\quad \begin{array}{r} \square \\ \div\ 7 \\ \hline 12 \end{array}$ $\quad \begin{array}{r} 84 \\ \div\ 12 \\ \hline \square \end{array}$ $\quad \begin{array}{r} \square \\ \div\ 12 \\ \hline 7 \end{array}$ $\quad \begin{array}{r} 84 \\ \div\ \square \\ \hline 12 \end{array}$

$7\overline{)84}$ with quotient $\square$ $\qquad \dfrac{84}{12} = \square$ $\qquad 12\overline{)84}$ with quotient $\square$ $\qquad \dfrac{84}{7} = \square$ $\qquad 7\overline{)84}$ with quotient $\square$

Activities

We have eighty-four pieces of dog food.

There are seven dog bowls.

When divided evenly, how many pieces of dog food will be in each bowl? ______

Use the numbers below to make equations.

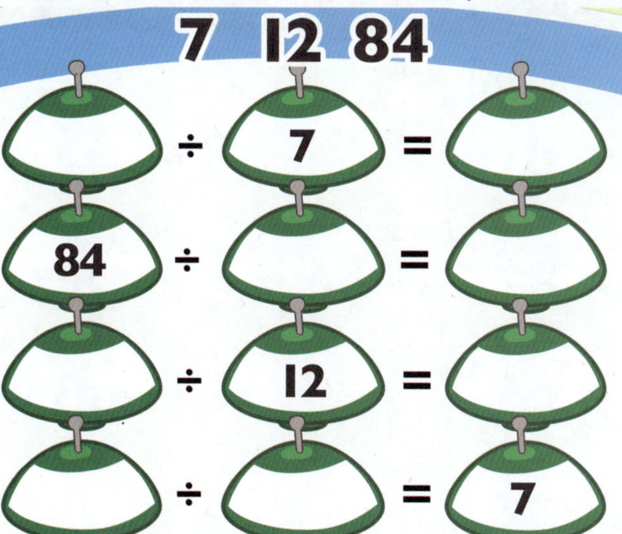

Make equations with the numbers below.

Divide the planets into twelve even groups by circling each group.

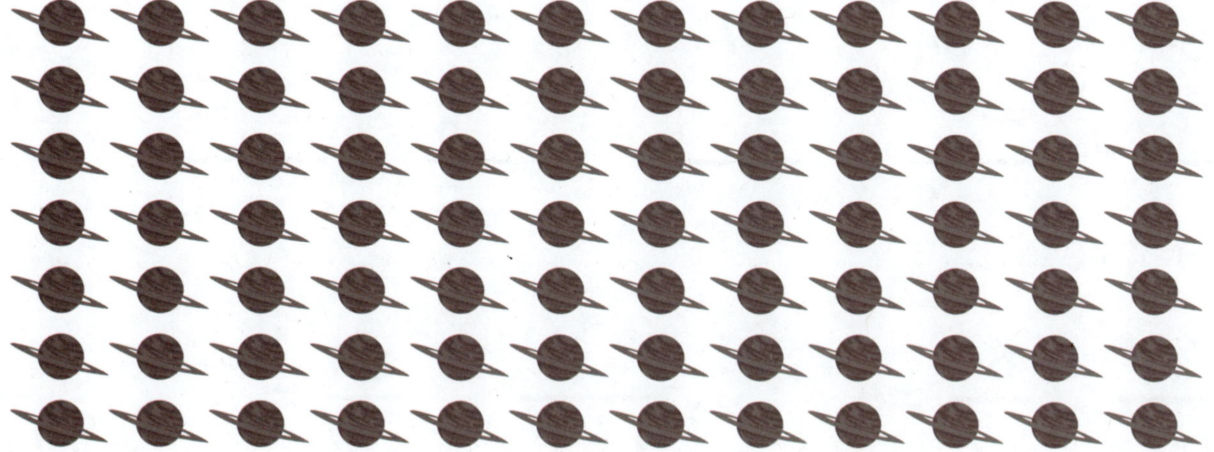

Write the equation to match the picture. ______ ÷ ______ = ______

Fill in the Answer

Fill in the blanks to complete the equations.

$64 \div \square = 8$ $\square \div 8 = 8$ $64 \div 8 = \square$

$64 \div 8 = \square$ $64 \div \square = 8$ $64 \div \square = 8$

$64 \div \square = 8$ $\square \div 8 = 8$ $\square \div 8 = 8$

$\square \div 8 = 8$ $64 \div 8 = \square$ $64 \div 8 = \square$

$\begin{array}{r} 64 \\ \div \; \square \\ \hline 8 \end{array}$ $\begin{array}{r} 64 \\ \div \; \square \\ \hline 8 \end{array}$ $\begin{array}{r} 64 \\ \div \; 8 \\ \hline \square \end{array}$ $\begin{array}{r} \square \\ \div \; 8 \\ \hline 8 \end{array}$ $\begin{array}{r} \square \\ \div \; 8 \\ \hline 8 \end{array}$ $\begin{array}{r} 64 \\ \div \; 8 \\ \hline \square \end{array}$

$8 \overline{)64}$ (quotient: $\square$) $\frac{64}{8} = \square$ $8 \overline{)64}$ (quotient: $\square$) $\frac{64}{8} = \square$ $8 \overline{)64}$ (quotient: $\square$)

Activities

There are sixty-four beakers.

There are eight shelves.

When divided evenly, how many beakers do we put on each shelf? ______

Use the numbers below to make equations.

Make equations with the numbers below.

8 8 64

Divide the awards into eight even groups by circling each group.

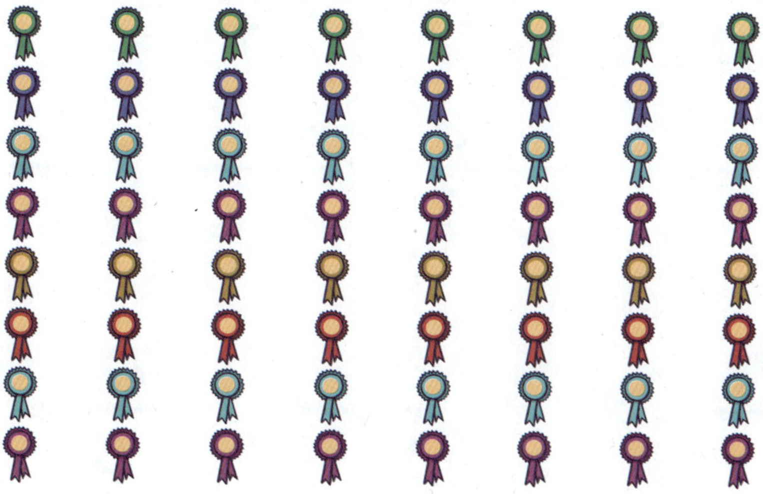

Write the equation to match the picture. ______ ÷ ______ = ______

Fill in the Answer

Fill in the blanks to complete the equations.

$\square \div 9 = 8$ $72 \div \square = 9$ $72 \div 9 = \square$

$72 \div \square = 8$ $\square \div 9 = 8$ $72 \div \square = 9$

$72 \div 9 = \square$ $72 \div 8 = \square$ $\square \div 9 = 8$

$\square \div 8 = 9$ $72 \div \square = 8$ $72 \div 8 = \square$

$\begin{array}{r} 72 \\ \div \square \\ \hline 9 \end{array}$ $\begin{array}{r} 72 \\ \div\ 8 \\ \hline \square \end{array}$ $\begin{array}{r} \square \\ \div\ 8 \\ \hline 9 \end{array}$ $\begin{array}{r} 72 \\ \div\ 9 \\ \hline \square \end{array}$ $\begin{array}{r} \square \\ \div\ 9 \\ \hline 8 \end{array}$ $\begin{array}{r} 72 \\ \div \square \\ \hline 8 \end{array}$

$\begin{array}{r} \square \\ 8\overline{)72} \end{array}$ $\frac{72}{9} = \square$ $\begin{array}{r} \square \\ 9\overline{)72} \end{array}$ $\frac{72}{8} = \square$ $\begin{array}{r} \square \\ 8\overline{)72} \end{array}$

Activities

There are seventy-two coins.

There are eight people. They share the coins evenly.

How many coins will each person get? ____

Use the numbers below to make equations.

8 9 72

___ ÷ ___ = 8

___ ÷ ___ = 9

___ ÷ 8 = ___

72 ÷ ___ = ___

Make equations with the numbers below.

8 9 72

Divide the coins into nine even groups by circling each group.

Write the equation to match the picture. ________ ÷ ________ = ________

Fill in the Answer

Fill in the blanks to complete the equations.

80 ÷ ___ = 10	80 ÷ 8 = ___	___ ÷ 8 = 10
___ ÷ 10 = 8	80 ÷ ___ = 10	80 ÷ 8 = ___
80 ÷ 10 = ___	___ ÷ 8 = 10	___ ÷ 10 = 8
80 ÷ ___ = 8	80 ÷ 10 = ___	80 ÷ ___ = 8

80	___	___	___	80	80
÷ 10	÷ 10	÷ 8	÷ 10	÷ ___	÷ 8
___	8	10	8	10	___

$8\overline{)80}$ = ___	$\frac{80}{10}$ = ___	$10\overline{)80}$ = ___	$\frac{80}{8}$ = ___	$8\overline{)80}$ = ___

Activities

There are eighty bananas.

There are eight bananas in each bunch.

How many bunches of bananas are there? _____

Use the numbers below to make equations.

8 10 80

☐ ÷ ☐ = 8

☐ ÷ ☐ = 10

☐ ÷ 10 = ☐

80 ÷ ☐ = ☐

Make equations with the numbers below.

8 10 80

Divide the keys into ten even groups by circling each group.

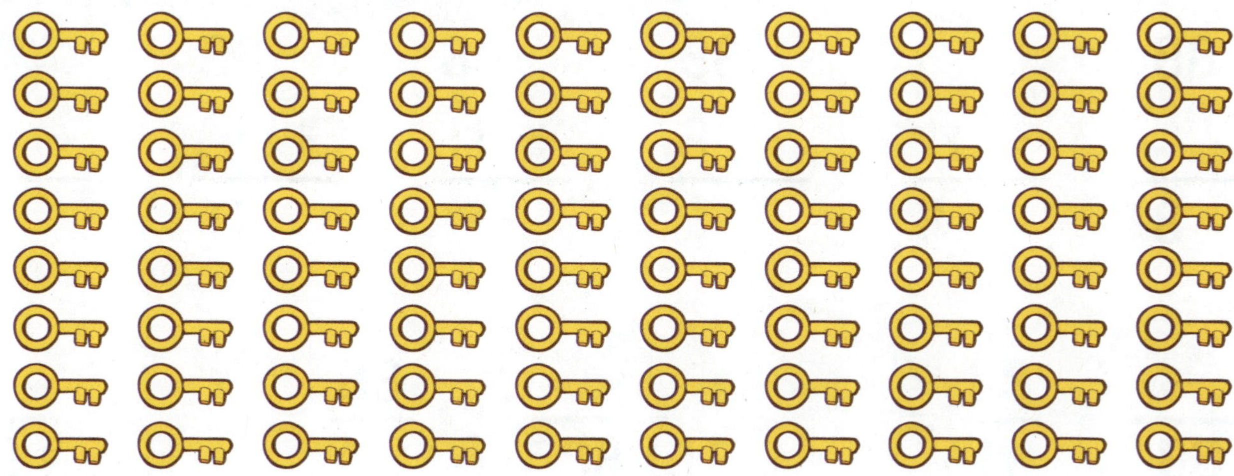

Write the equation to match the picture. _______ ÷ _______ = _______

Fill in the Answer

Fill in the blanks to complete the equations.

$88 \div 11 = \square$	$\square \div 8 = 11$	$88 \div \square = 8$
$88 \div \square = 11$	$88 \div 8 = \square$	$88 \div 11 = \square$
$\square \div 11 = 8$	$88 \div \square = 11$	$\square \div 8 = 11$
$88 \div \square = 8$	$\square \div 11 = 8$	$88 \div 8 = \square$

$$\begin{array}{r} 88 \\ \div\ \square \\ \hline 8 \end{array} \quad \begin{array}{r} \square \\ \div\ 11 \\ \hline 8 \end{array} \quad \begin{array}{r} 88 \\ \div\ 8 \\ \hline \square \end{array} \quad \begin{array}{r} 88 \\ \div\ 11 \\ \hline \square \end{array} \quad \begin{array}{r} 88 \\ \div\ \square \\ \hline 11 \end{array} \quad \begin{array}{r} \square \\ \div\ 8 \\ \hline 11 \end{array}$$

$$8\overline{)88}\ (\square) \qquad \frac{88}{11} = \square \qquad 11\overline{)88}\ (\square) \qquad \frac{88}{8} = \square \qquad 8\overline{)88}\ (\square)$$

Activities

There are eighty-eight birds.

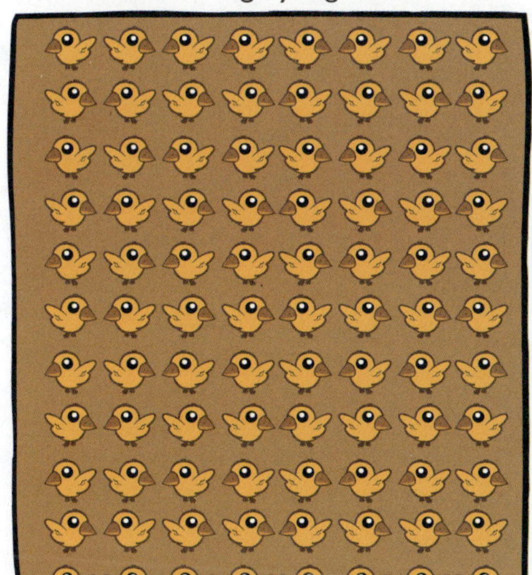

There are eleven wires.

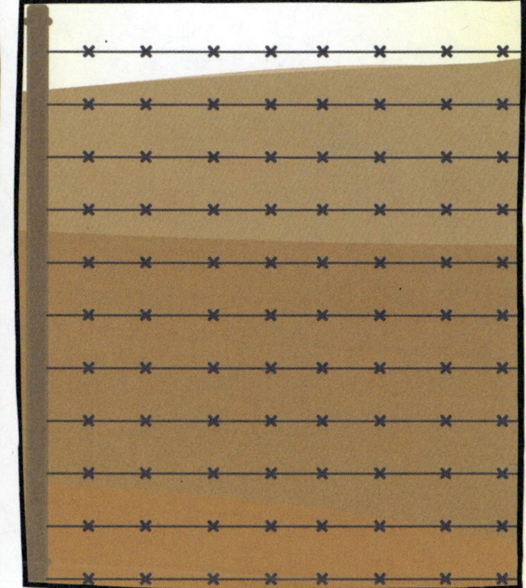

How many birds will sit on each wire when divided evenly? _____

Use the numbers below to make equations.

8 11 88

☐ ÷ ☐ = 8

☐ ÷ ☐ = 11

88 ÷ ☐ = ☐

☐ ÷ 8 = ☐

Make equations with the numbers below.

Divide the barrels into eight even groups by circling each group.

Write the equation to match the picture. ______ ÷ ______ = ______

Fill in the Answer

Fill in the blanks to complete the equations.

$96 \div \square = 12$	$96 \div 8 = \square$	$96 \div 12 = \square$
$96 \div 8 = \square$	$\square \div 8 = 12$	$\square \div 12 = 8$
$\square \div 12 = 8$	$96 \div \square = 12$	$96 \div \square = 8$
$96 \div \square = 8$	$96 \div 12 = \square$	$\square \div 8 = 12$

$$\begin{array}{r} 96 \\ \div\ 8 \\ \hline \square \end{array} \qquad \begin{array}{r} \square \\ \div\ 8 \\ \hline 12 \end{array} \qquad \begin{array}{r} 96 \\ \div\ 12 \\ \hline \square \end{array} \qquad \begin{array}{r} \square \\ \div\ 12 \\ \hline 8 \end{array} \qquad \begin{array}{r} 96 \\ \div\ 8 \\ \hline \square \end{array} \qquad \begin{array}{r} \square \\ \div\ 8 \\ \hline 12 \end{array}$$

$$\begin{array}{r} \square \\ 8\overline{)96} \end{array} \qquad \frac{96}{12} = \square \qquad \begin{array}{r} \square \\ 12\overline{)96} \end{array} \qquad \frac{96}{8} = \square \qquad \begin{array}{r} \square \\ 8\overline{)96} \end{array}$$

Activities

There are ninety-six snowballs.

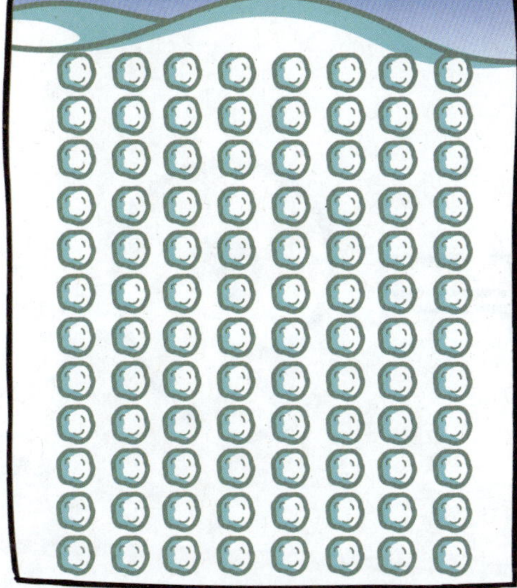

There are twelve snowballs in each group.

How many groups of snowballs do we have in all? ______

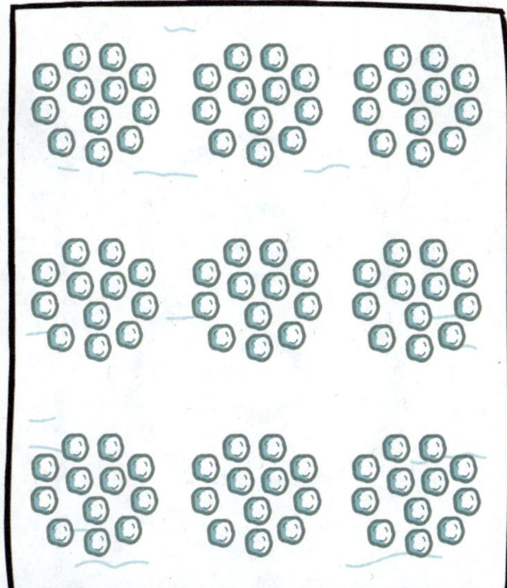

Use the numbers below to make equations.

8 12 96

96	÷		=	
	÷		=	12
	÷		=	8
	÷	12	=	

Make equations with the numbers below.

8 12 96

Divide the igloos into eight even groups by circling each group.

Write the equation to match the picture. ______ ÷ ______ = ______

Fill in the Answer

Fill in the blanks to complete the equations.

$\square \div 9 = 9$ $\quad$ $81 \div 9 = \square$ $\quad$ $81 \div 9 = \square$

$81 \div \square = 9$ $\quad$ $\square \div 9 = 9$ $\quad$ $81 \div 9 = \square$

$\square \div 9 = 9$ $\quad$ $81 \div 9 = \square$ $\quad$ $\square \div 9 = 9$

$81 \div \square = 9$ $\quad$ $\square \div 9 = 9$ $\quad$ $81 \div \square = 9$

$$\begin{array}{r} 81 \\ \div\ 9 \\ \hline \square \end{array} \quad \begin{array}{r} \square \\ \div\ 9 \\ \hline 9 \end{array} \quad \begin{array}{r} 81 \\ \div\ \square \\ \hline 9 \end{array} \quad \begin{array}{r} 81 \\ \div\ 9 \\ \hline \square \end{array} \quad \begin{array}{r} 81 \\ \div\ \square \\ \hline 9 \end{array} \quad \begin{array}{r} \square \\ \div\ 9 \\ \hline 9 \end{array}$$

$$9\overset{\square}{\overline{)81}} \quad \frac{81}{9} = \square \quad 9\overset{\square}{\overline{)81}} \quad \frac{81}{9} = \square \quad 9\overset{\square}{\overline{)81}}$$

Activities

There are eighty-one drum sticks.

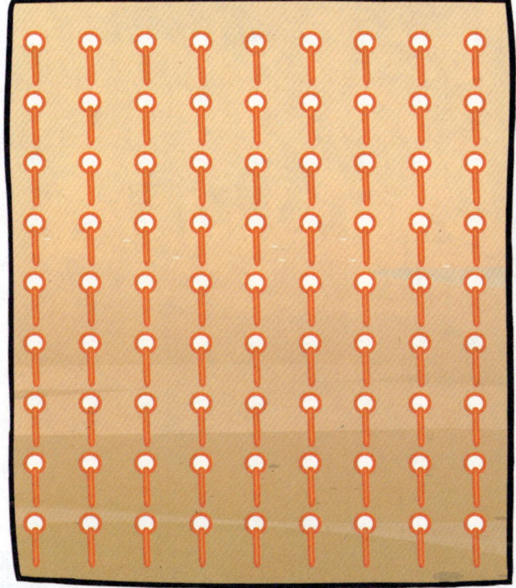

Nine sticks fit into one bucket.

How many buckets do we need to hold all of the sticks? ______

Use the numbers below to make equations.

9 9 81

___ ÷ ___ = 9

___ ÷ 9 = ___

___ ÷ ___ = 9

81 ÷ ___ = ___

Make equations with the numbers below.

9 9 81

Divide the drumsticks into nine even groups by circling each group.

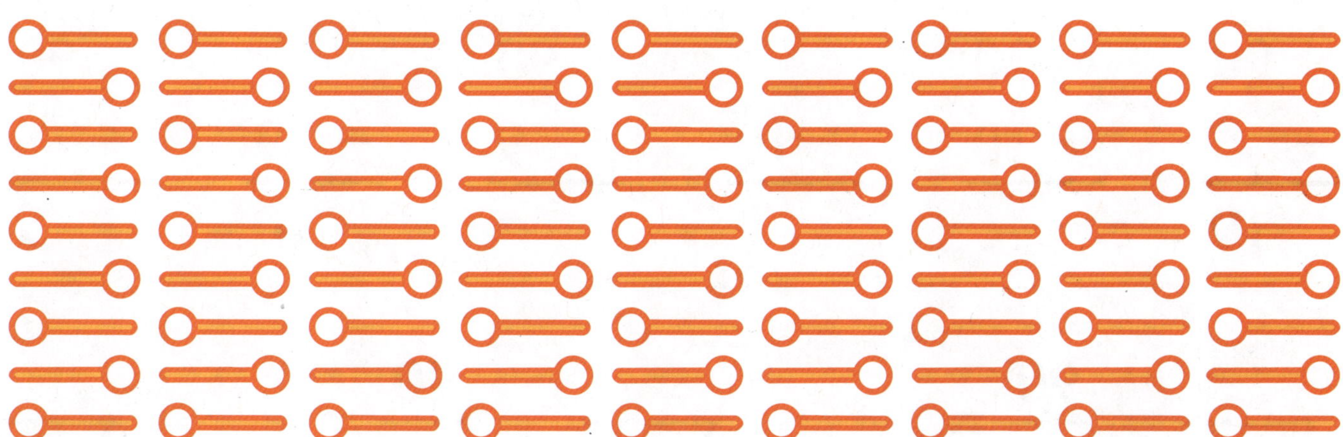

Write the equation to match the picture. ______ ÷ ______ = ______

Fill in the Answer

Fill in the blanks to complete the equations.

$90 \div 10 = \square$	$90 \div \square = 9$	$\square \div 10 = 9$
$\square \div 10 = 9$	$\square \div 9 = 10$	$90 \div \square = 9$
$90 \div \square = 10$	$90 \div 10 = \square$	$90 \div 9 = \square$
$90 \div 9 = \square$	$90 \div \square = 10$	$\square \div 9 = 10$

$$\begin{array}{r} 90 \\ \div\ \square \\ \hline 10 \end{array} \quad \begin{array}{r} 90 \\ \div\ 10 \\ \hline \square \end{array} \quad \begin{array}{r} \square \\ \div\ 10 \\ \hline 9 \end{array} \quad \begin{array}{r} 90 \\ \div\ 9 \\ \hline \square \end{array} \quad \begin{array}{r} \square \\ \div\ 9 \\ \hline 10 \end{array} \quad \begin{array}{r} 90 \\ \div\ \square \\ \hline 9 \end{array}$$

$$9\overline{)90}\ (\square) \quad \frac{90}{9} = \square \quad 10\overline{)90}\ (\square) \quad \frac{90}{10} = \square \quad 9\overline{)90}\ (\square)$$

Activities

There are ninety bubbles.

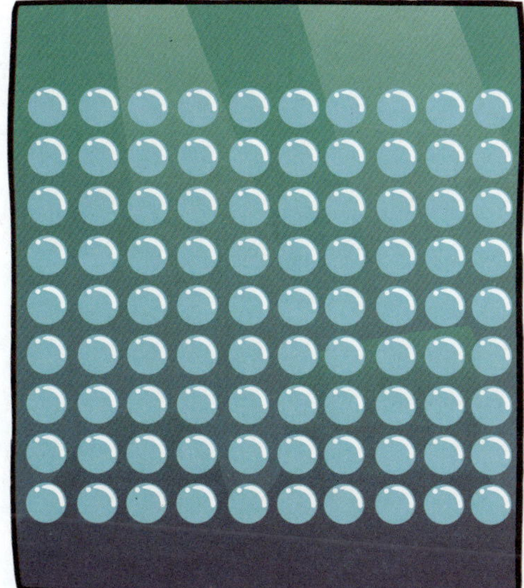

There are ten bubbles in each group.

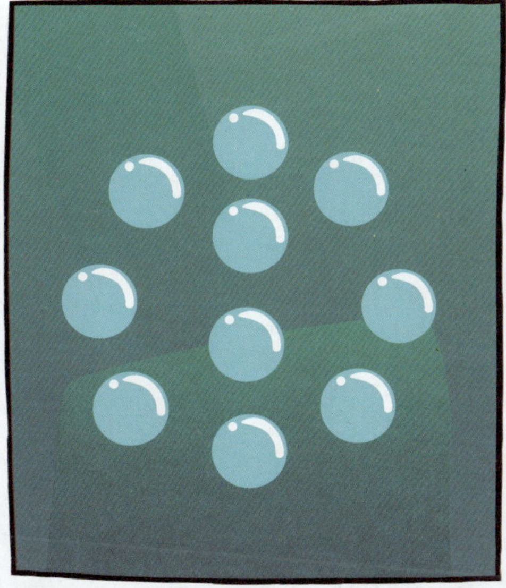

How many groups of bubbles are there in all? _____

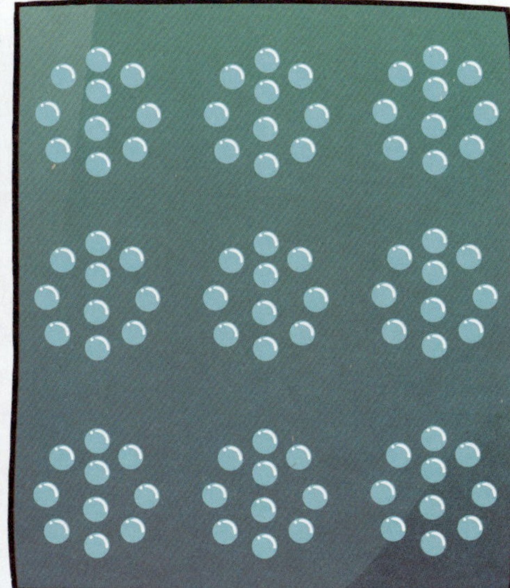

Use the numbers below to make equations.

9 10 90

$\_\_\_ \div \_\_\_ = 9$

$\_\_\_ \div 10 = \_\_\_$

$\_\_\_ \div \_\_\_ = 10$

$90 \div \_\_\_ = \_\_\_$

Make equations with the numbers below.

9 10 90

Divide the turtles into nine even groups by circling each group.

Write the equation to match the picture. _______ ÷ _______ = _______

Fill in the Answer

Fill in the blanks to complete the equations.

$99 \div \square = 11$ $\square \div 11 = 9$ $\square \div 9 = 11$

$99 \div 11 = \square$ $99 \div 9 = \square$ $99 \div 9 = \square$

$\square \div 9 = 11$ $99 \div \square = 11$ $\square \div 11 = 9$

$99 \div \square = 9$ $99 \div 11 = \square$ $99 \div \square = 9$

$$\begin{array}{r}\square\\ \div\ 9\\ \hline 11\end{array}\quad \begin{array}{r}99\\ \div\ 9\\ \hline \square\end{array}\quad \begin{array}{r}99\\ \div\ \square\\ \hline 9\end{array}\quad \begin{array}{r}\square\\ \div\ 11\\ \hline 9\end{array}\quad \begin{array}{r}99\\ \div\ \square\\ \hline 11\end{array}\quad \begin{array}{r}99\\ \div\ 11\\ \hline \square\end{array}$$

$$\begin{array}{r}\square\\ 9\overline{)99}\end{array}\quad \frac{99}{9} = \square \quad \begin{array}{r}\square\\ 11\overline{)99}\end{array}\quad \frac{99}{11} = \square \quad \begin{array}{r}\square\\ 9\overline{)99}\end{array}$$

Activities

There are ninety-nine bones.

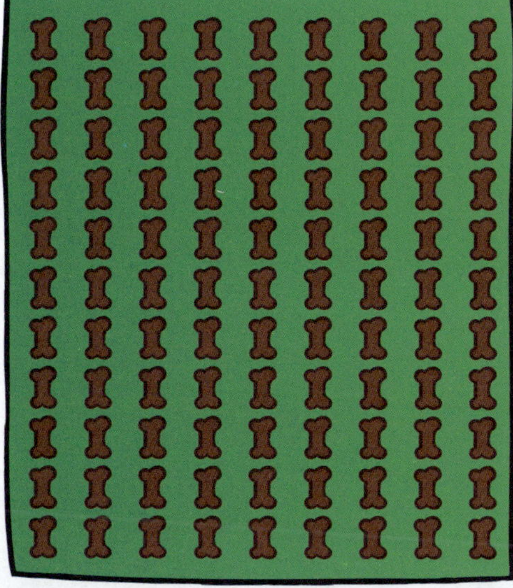

We will put eleven bones in each bowl.

How many bowls will we need? _____

Use the numbers below to make equations.

9 11 99

99 ÷ ___ = ___

___ ÷ ___ = 11

___ ÷ 11 = ___

___ ÷ ___ = 9

Make equations with the numbers below.

9 11 99

Divide the dog bones into nine even groups by circling each group.

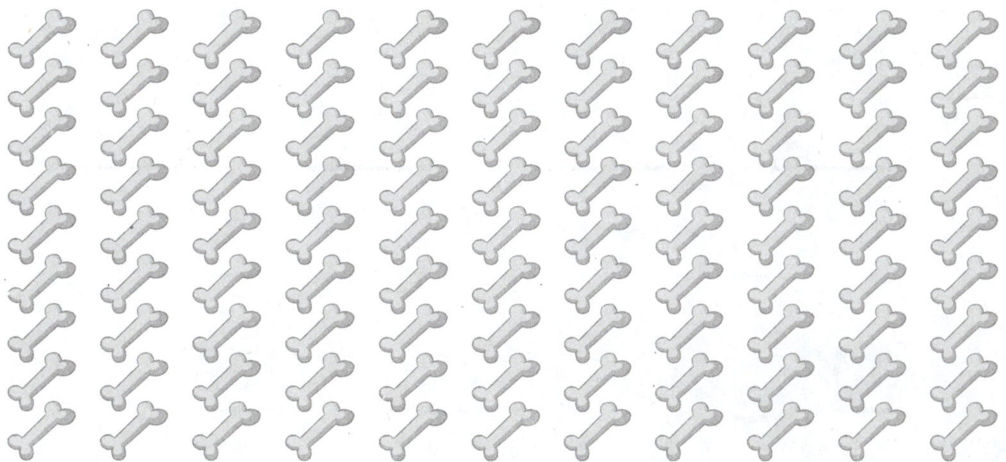

Write the equation to match the picture. ______ ÷ ______ = ______

Fill in the Answer

Fill in the blanks to complete the equations.

$108 \div 9 = \_\_\_$ $\quad$ $\_\_\_ \div 9 = 12$ $\quad$ $\_\_\_ \div 12 = 9$

$108 \div 12 = \_\_\_$ $\quad$ $108 \div \_\_\_ = 9$ $\quad$ $108 \div \_\_\_ = 12$

$108 \div \_\_\_ = 9$ $\quad$ $\_\_\_ \div 9 = 12$ $\quad$ $108 \div 12 = \_\_\_$

$\_\_\_ \div 9 = 12$ $\quad$ $108 \div \_\_\_ = 9$ $\quad$ $108 \div 9 = \_\_\_$

$$\begin{array}{r} \_\_\_ \\ \div\ 12 \\ \hline 9 \end{array} \qquad \begin{array}{r} 108 \\ \div\ 9 \\ \hline \_\_\_ \end{array} \qquad \begin{array}{r} \_\_\_ \\ \div\ 9 \\ \hline 12 \end{array} \qquad \begin{array}{r} \_\_\_ \\ \div\ 12 \\ \hline 9 \end{array} \qquad \begin{array}{r} 108 \\ \div\ 12 \\ \hline \_\_\_ \end{array} \qquad \begin{array}{r} 108 \\ \div\ \_\_\_ \\ \hline 9 \end{array}$$

$$9\overline{)108}\ \text{quotient: } \_\_\_ \qquad \frac{108}{9} = \_\_\_ \qquad 12\overline{)108}\ \text{quotient: } \_\_\_ \qquad \frac{108}{12} = \_\_\_ \qquad 9\overline{)108}\ \text{quotient: } \_\_\_$$

Activities

There are one hundred and eight fish.

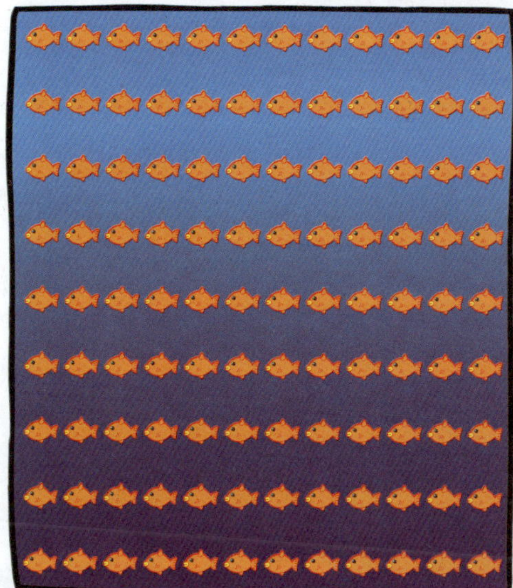

They gather in groups of twelve.

When divided evenly, how many groups of fish are there? ______

Use the numbers below to make equations.

9 12 108

108 ÷ ___ = ___

___ ÷ ___ = 12

___ ÷ ___ = 9

___ ÷ 12 = ___

Make equations with the numbers below.

9 12 108

Divide the fish into twelve even groups by circling each group.

Write the equation to match the picture. ________ ÷ ________ = ________

Fill in the Answer

Fill in the blanks to complete the equations.

$\square \div 10 = 10$ $100 \div 10 = \square$ $100 \div 10 = \square$

$100 \div 10 = \square$ $\square \div 10 = 10$ $\square \div 10 = 10$

$100 \div \square = 10$ $100 \div \square = 10$ $100 \div \square = 10$

$100 \div 10 = \square$ $\square \div 10 = 10$ $100 \div \square = 10$

$\begin{array}{r} 100 \\ \div\ 10 \\ \hline \square \end{array}$ $\begin{array}{r} \square \\ \div\ 10 \\ \hline 10 \end{array}$ $\begin{array}{r} 100 \\ \div\ \square \\ \hline 10 \end{array}$ $\begin{array}{r} 100 \\ \div\ 10 \\ \hline \square \end{array}$ $\begin{array}{r} 100 \\ \div\ \square \\ \hline 10 \end{array}$ $\begin{array}{r} \square \\ \div\ 10 \\ \hline 10 \end{array}$

$10\overline{)100}$ with $\square$ above; $\frac{100}{10} = \square$; $10\overline{)100}$ with $\square$ above; $\frac{100}{10} = \square$; $10\overline{)100}$ with $\square$ above

Activities

There are one hundred tires.

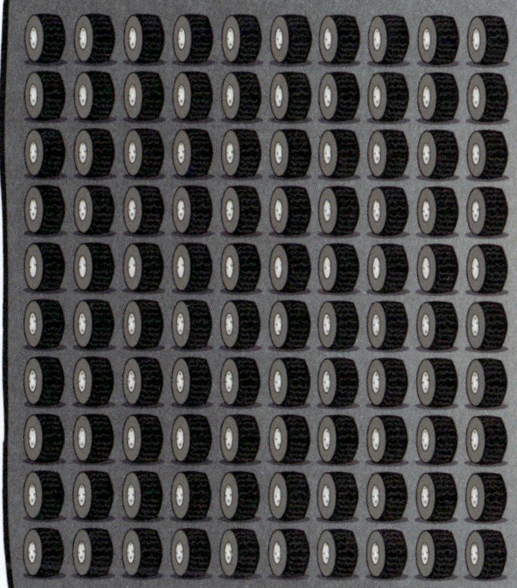

We will stack them into groups of ten.

How many stacks of tires are there when divided evenly? ______

Divide the flags into ten even groups by circling each group.

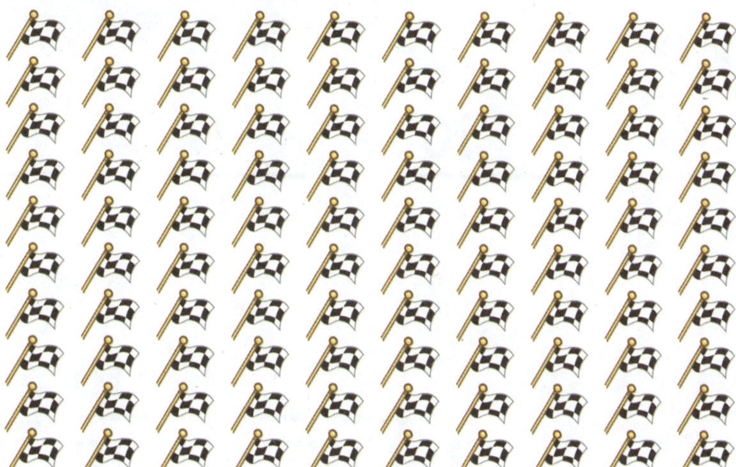

Write the equation to match the picture. ______ ÷ ______ = ______

Fill in the Answer

Fill in the blanks to complete the equations.

$\square \div 10 = 11$ $110 \div 10 = \square$ $\square \div 10 = 11$

$110 \div 11 = \square$ $110 \div \square = 11$ $110 \div 10 = \square$

$110 \div \square = 11$ $\square \div 11 = 10$ $\square \div 11 = 10$

$110 \div \square = 10$ $110 \div 11 = \square$ $110 \div \square = 10$

$\begin{array}{r} \square \\ \div\ 11 \\ \hline 10 \end{array}$ $\begin{array}{r} 110 \\ \div\ 10 \\ \hline \square \end{array}$ $\begin{array}{r} \square \\ \div\ 10 \\ \hline 11 \end{array}$ $\begin{array}{r} \square \\ \div\ 11 \\ \hline 10 \end{array}$ $\begin{array}{r} 110 \\ \div\ 11 \\ \hline \square \end{array}$ $\begin{array}{r} 110 \\ \div\ \square \\ \hline 10 \end{array}$

$10\overline{)110}$ = $\square$ $\frac{110}{11} = \square$ $11\overline{)110}$ = $\square$ $\frac{110}{10} = \square$ $10\overline{)110}$ = $\square$

Activities

There are one hundred and ten seeds.

There are ten birds.

When divided evenly, how many seeds does each bird get? _____

Use the numbers below to make equations.

Make equations with the numbers below.

Divide the birds into eleven even groups by circling each group.

Write the equation to match the picture. _______ ÷ _______ = _______

Fill in the Answer

Fill in the blanks to complete the equations.

$120 \div \square = 12$ $\quad$ $120 \div 12 = \square$ $\quad$ $120 \div \square = 12$

$\square \div 10 = 12$ $\quad$ $\square \div 10 = 12$ $\quad$ $120 \div 12 = \square$

$120 \div 10 = \square$ $\quad$ $120 \div 10 = \square$ $\quad$ $120 \div \square = 10$

$\square \div 12 = 10$ $\quad$ $120 \div \square = 10$ $\quad$ $\square \div 12 = 10$

$$\begin{array}{r} \square \\ \div\ 12 \\ \hline 10 \end{array} \quad \begin{array}{r} 120 \\ \div\ 10 \\ \hline \square \end{array} \quad \begin{array}{r} \square \\ \div\ 10 \\ \hline 12 \end{array} \quad \begin{array}{r} \square \\ \div\ 12 \\ \hline 10 \end{array} \quad \begin{array}{r} 120 \\ \div\ 12 \\ \hline \square \end{array} \quad \begin{array}{r} 120 \\ \div\ \square \\ \hline 12 \end{array}$$

$$\begin{array}{r} \square \\ 10\overline{)120} \end{array} \quad \frac{120}{10} = \square \quad \begin{array}{r} \square \\ 12\overline{)120} \end{array} \quad \frac{120}{12} = \square \quad \begin{array}{r} \square \\ 10\overline{)120} \end{array}$$

Activities

There are one hundred and twenty fireflies.

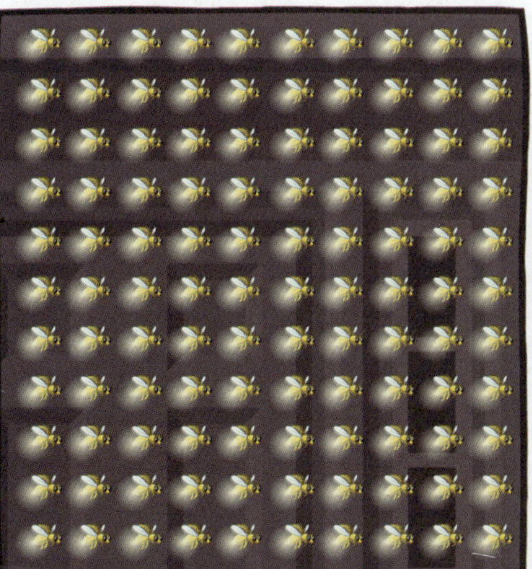

They gather in groups of ten.

How many groups of fireflies are there when divided evenly? _____

Divide the fireflies into twelve even groups by circling each group.

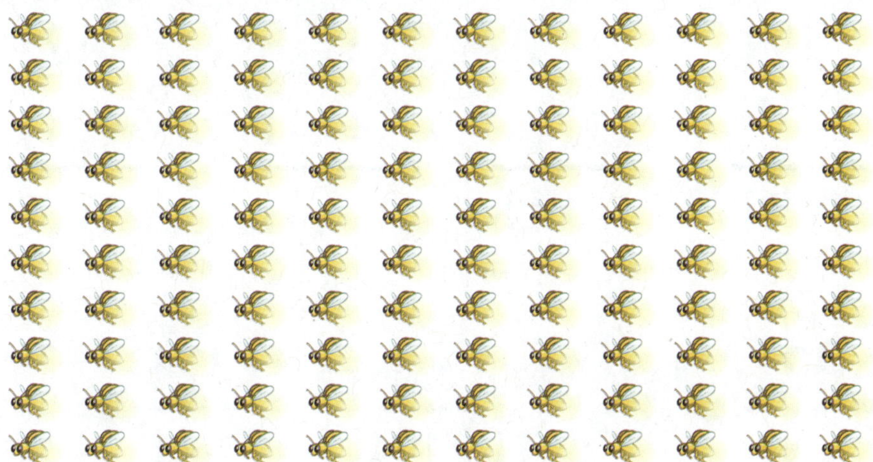

Write the equation to match the picture. _______ ÷ _______ = _______

Fill in the Answer

Fill in the blanks to complete the equations.

$121 \div \square = 11$ $121 \div 11 = \square$ $121 \div 11 = \square$

$\square \div 11 = 11$ $121 \div \square = 11$ $121 \div \square = 11$

$121 \div \square = 11$ $\square \div 11 = 11$ $\square \div 11 = 11$

$\square \div 11 = 11$ $121 \div 11 = \square$ $121 \div 11 = \square$

$$\begin{array}{r} 121 \\ \div\ 11 \\ \hline \square \end{array} \quad \begin{array}{r} 121 \\ \div\ \square \\ \hline 11 \end{array} \quad \begin{array}{r} \square \\ \div\ 11 \\ \hline 11 \end{array} \quad \begin{array}{r} \square \\ \div\ 11 \\ \hline 11 \end{array} \quad \begin{array}{r} 121 \\ \div\ 11 \\ \hline \square \end{array} \quad \begin{array}{r} \square \\ \div\ 11 \\ \hline 11 \end{array}$$

$$\begin{array}{r} \square \\ 11\overline{)121} \end{array} \quad \frac{121}{11} = \square \quad \begin{array}{r} \square \\ 11\overline{)121} \end{array} \quad \frac{121}{11} = \square \quad \begin{array}{r} \square \\ 11\overline{)121} \end{array}$$

Activities

There are one hundred and twenty-one snowballs.

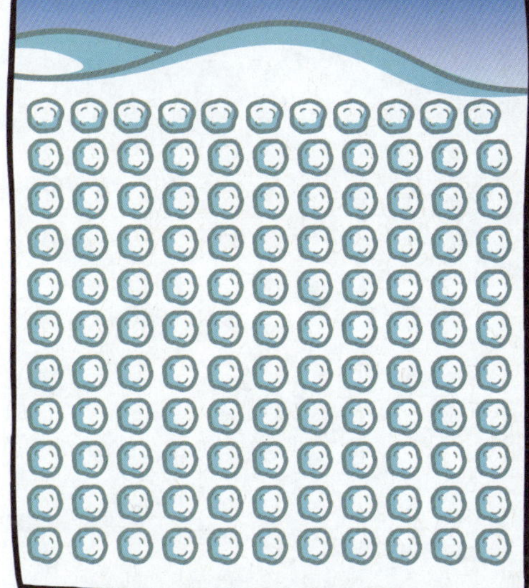

There are eleven buckets.

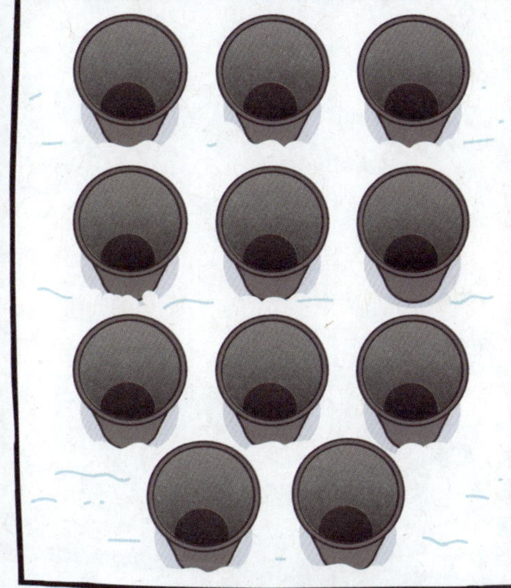

When divided evenly, how many snowballs are in each bucket? _____

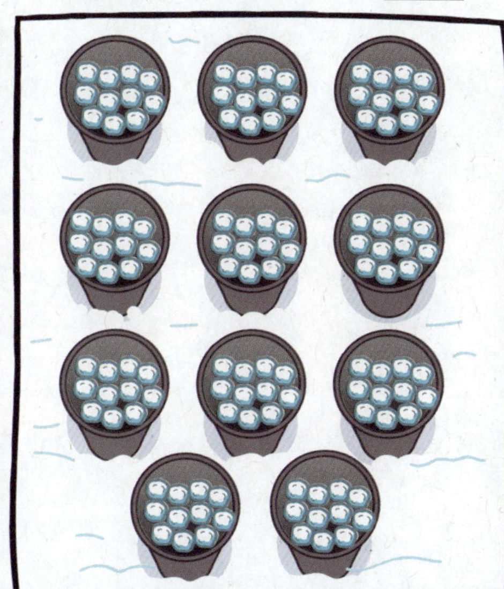

Use the numbers below to make equations.

11 11 121

121 ÷ ___ = ___

___ ÷ 11 = ___

___ ÷ ___ = 11

121 ÷ ___ = ___

Make equations with the numbers below.

11 11 121

Divide the snowballs into eleven even groups by circling each group.

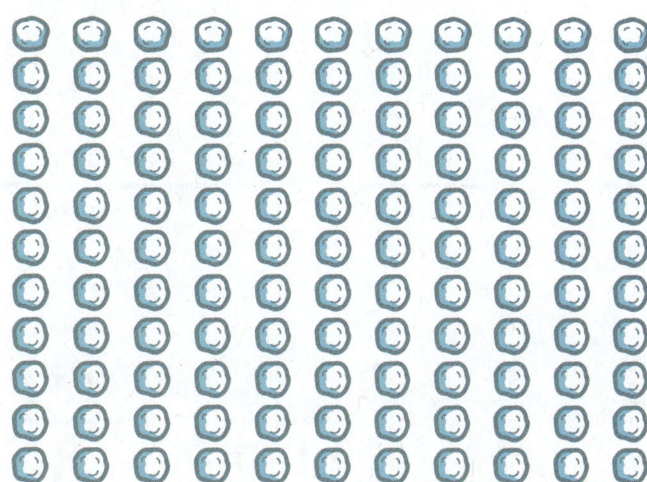

Write the equation to match the picture. _____ ÷ _____ = _____

Fill in the Answer

Fill in the blanks to complete the equations.

$132 \div 11 = \square$ $\quad$ $132 \div \square = 11$ $\quad$ $132 \div \square = 12$

$132 \div \square = 12$ $\quad$ $\square \div 12 = 11$ $\quad$ $132 \div 12 = \square$

$\square \div 11 = 12$ $\quad$ $132 \div 11 = \square$ $\quad$ $\square \div 12 = 11$

$132 \div 12 = \square$ $\quad$ $\square \div 11 = 12$ $\quad$ $132 \div \square = 11$

$\begin{array}{r} 132 \\ \div\ \square \\ \hline 12 \end{array}$ $\quad$ $\begin{array}{r} 132 \\ \div\ 11 \\ \hline \square \end{array}$ $\quad$ $\begin{array}{r} \square \\ \div\ 11 \\ \hline 12 \end{array}$ $\quad$ $\begin{array}{r} 132 \\ \div\ 12 \\ \hline \square \end{array}$ $\quad$ $\begin{array}{r} \square \\ \div\ 12 \\ \hline 11 \end{array}$ $\quad$ $\begin{array}{r} 132 \\ \div\ \square \\ \hline 11 \end{array}$

$\begin{array}{r} \square \\ 11\overline{)132} \end{array}$ $\quad$ $\frac{132}{11} = \square$ $\quad$ $\begin{array}{r} \square \\ 12\overline{)132} \end{array}$ $\quad$ $\frac{132}{12} = \square$ $\quad$ $\begin{array}{r} \square \\ 11\overline{)132} \end{array}$

Activities

There are one hundred and thirty-two lamps.

There are eleven ropes.

How many lamps are on each rope when divided evenly? _____

Use the numbers below to make equations.

11 12 132

___ ÷ 11 = ___

___ ÷ ___ = 11

___ ÷ ___ = 12

132 ÷ ___ = ___

Make equations with the numbers below.

11 12 132

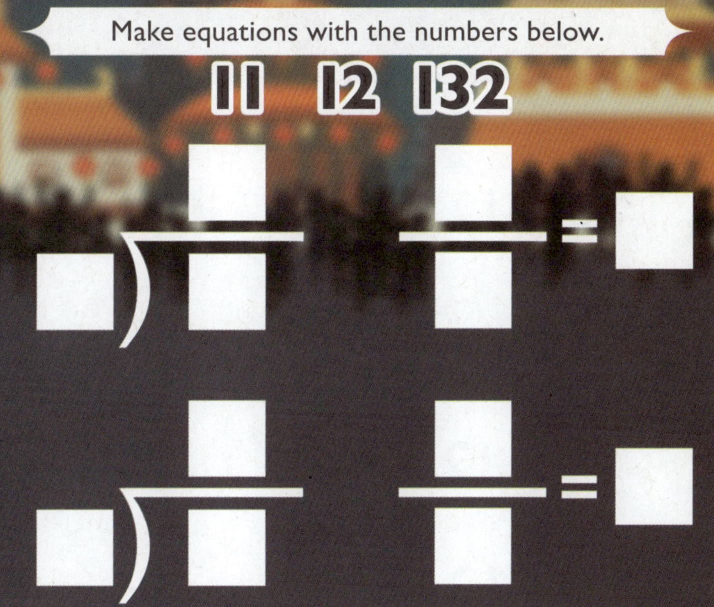

Divide the lamps into twelve even groups by circling each group.

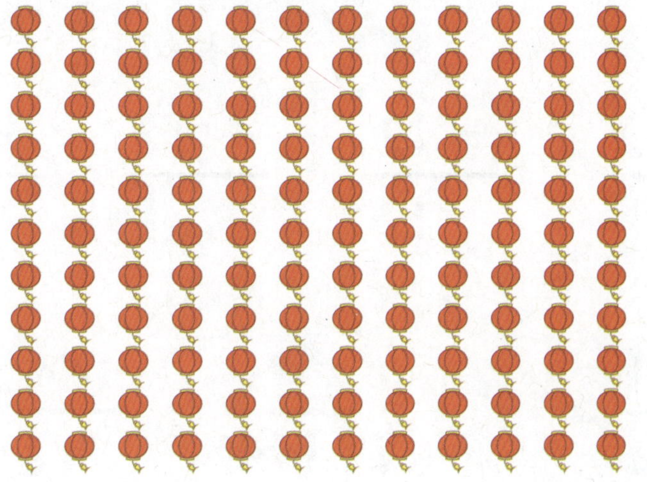

Write the equation to match the picture. _______ ÷ _______ = _______

Fill in the Answer

Fill in the blanks to complete the equations.

$144 \div 12 = \square$ $144 \div \square = 12$ $144 \div \square = 12$

$144 \div \square = 12$ $\square \div 12 = 12$ $144 \div 12 = \square$

$\square \div 12 = 12$ $144 \div 12 = \square$ $\square \div 12 = 12$

$144 \div 12 = \square$ $\square \div 12 = 12$ $144 \div \square = 12$

$\begin{array}{r} 144 \\ \div \square \\ \hline 12 \end{array}$ $\begin{array}{r} 144 \\ \div 12 \\ \hline \square \end{array}$ $\begin{array}{r} \square \\ \div 12 \\ \hline 12 \end{array}$ $\begin{array}{r} 144 \\ \div 12 \\ \hline \square \end{array}$ $\begin{array}{r} \square \\ \div 12 \\ \hline 12 \end{array}$ $\begin{array}{r} 144 \\ \div \square \\ \hline 12 \end{array}$

$12\overline{)144}$ (answer: $\square$) $\frac{144}{12} = \square$ $12\overline{)144}$ (answer: $\square$) $\frac{144}{12} = \square$ $12\overline{)144}$ (answer: $\square$)

Activities

There are one hundred and forty-four shark fins.

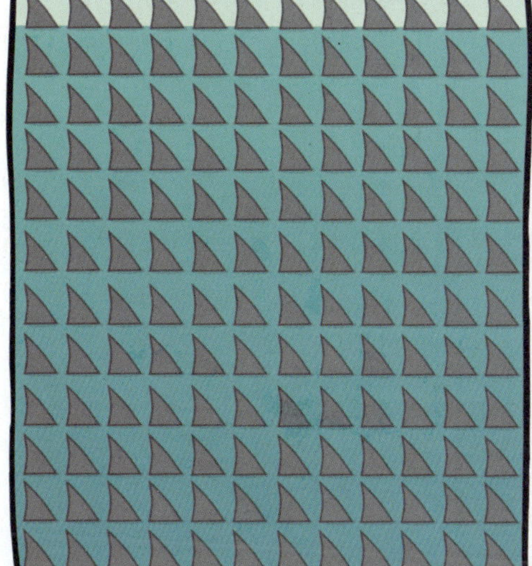

There are twelve islands.

When divided evenly, how many sharks swim around each island? ______

Use the numbers below to make equations.

12 12 144

144 ÷ ___ = ___

___ ÷ ___ = 12

___ ÷ ___ = 12

___ ÷ 12 = ___

Make equations with the numbers below.

12 12 144

Divide the stars into twelve even groups by circling each group.

Write the equation to match the picture. ______ ÷ ______ = ______

Activities

Activities

Use the numbers below to make equations.

0 0 9

0 ÷ ___ = ___

___ ÷ ___ = 0

___ ÷ 9 = ___

0 ÷ ___ = ___

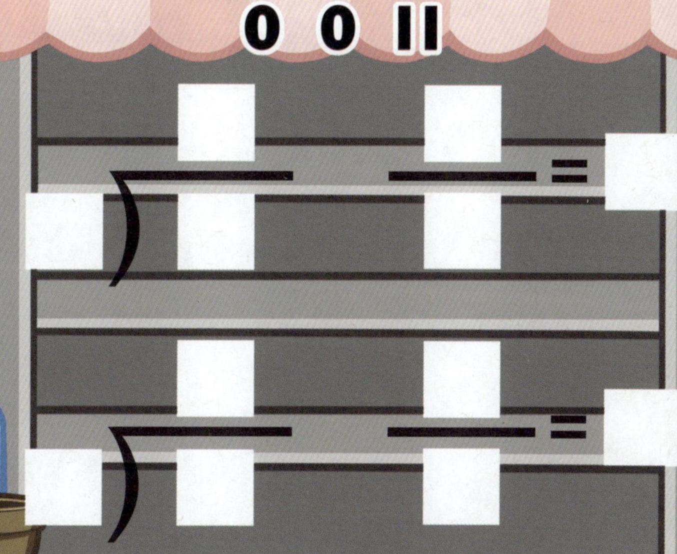

Activities

Activities

Activities

Activities

Activities

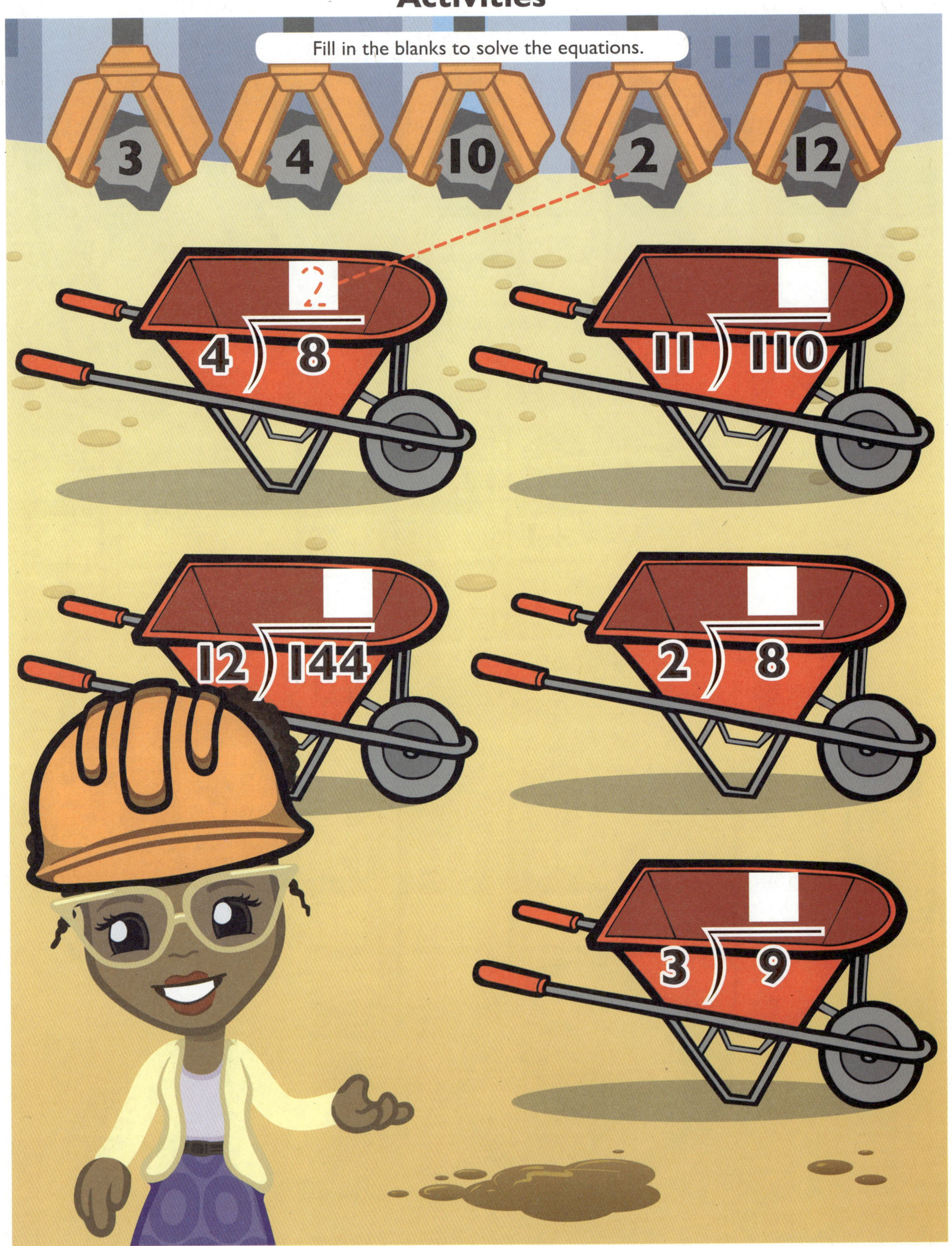

Activities

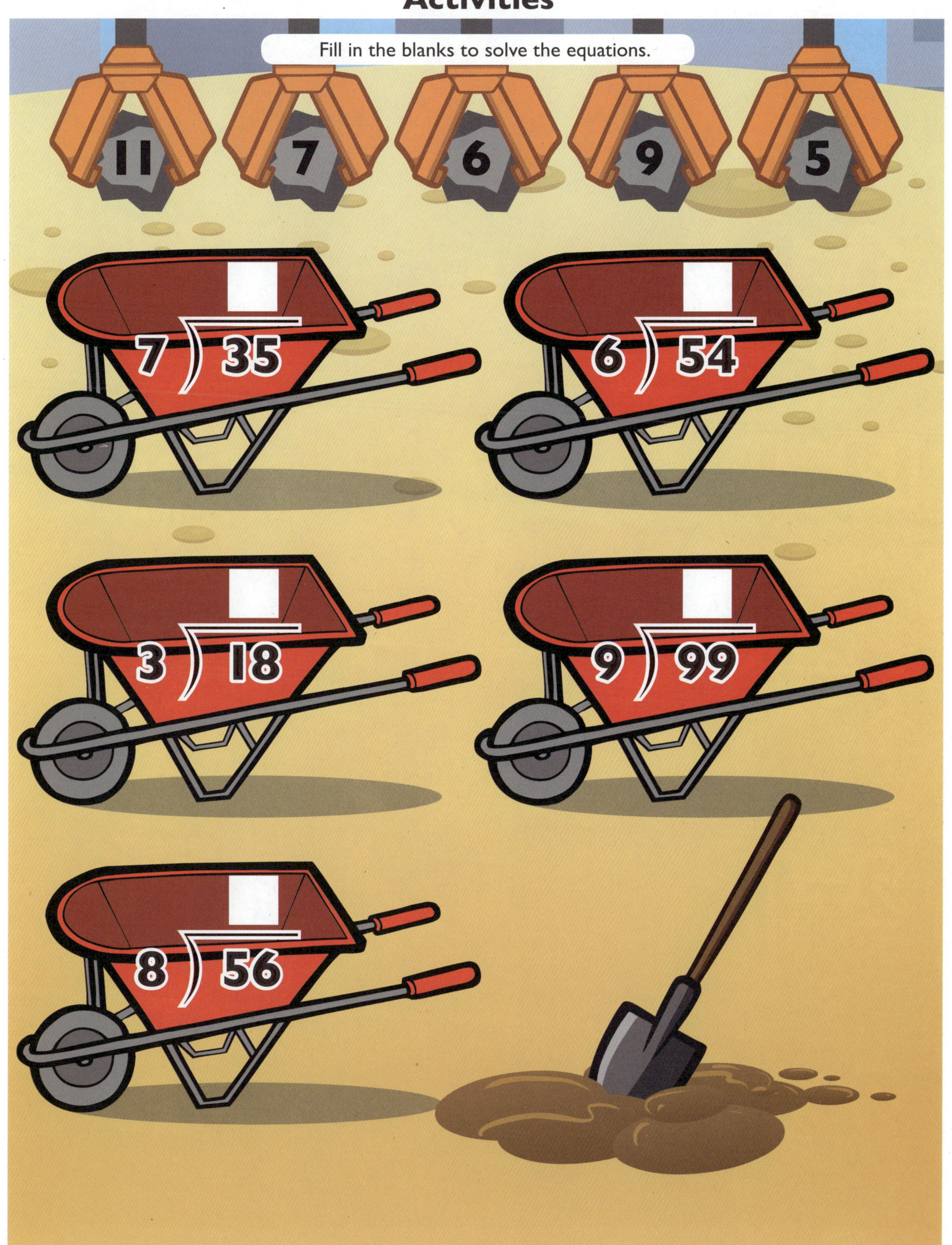

Divide by 1

Solve each problem and draw a line to connect the candle to the cupcake.

$4 \div 1$	7
$12 \div 1$	2
$7 \div 1$	4
$2 \div 1$	9
$9 \div 1$	12

Divide by 1

Divide by 2

Solve each problem and draw a line to connect the duck to the bathtub.
4÷2
12÷2
24÷2
16÷2
20÷2
12
10
6
2
8

Divide by 2

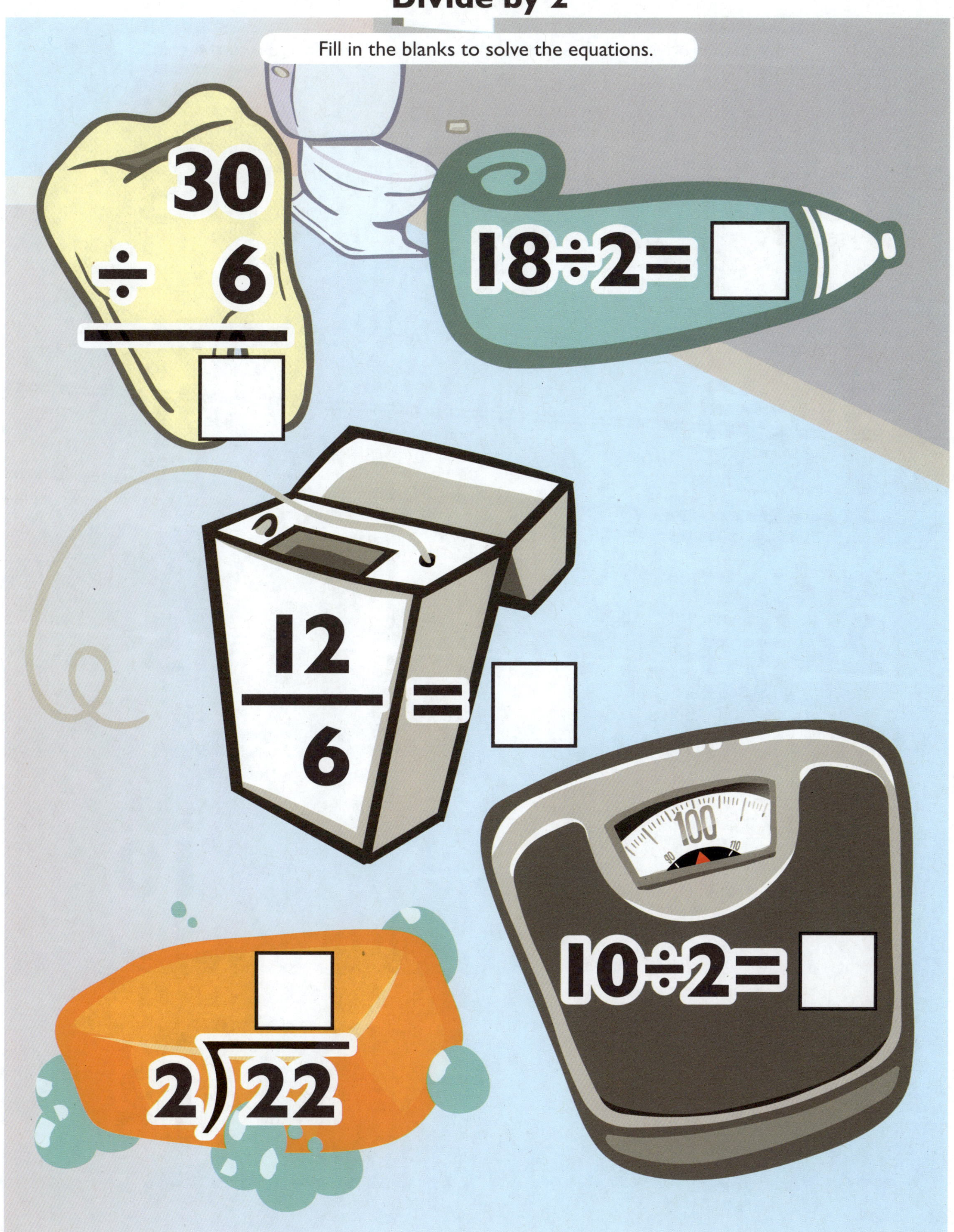

Divide by 3

Solve each problem and draw a line to connect the wheelbarrow to the gold.

$3 \div 3$	8
$30 \div 3$	1
$24 \div 3$	5
$36 \div 3$	10
$15 \div 3$	12

Divide by 3

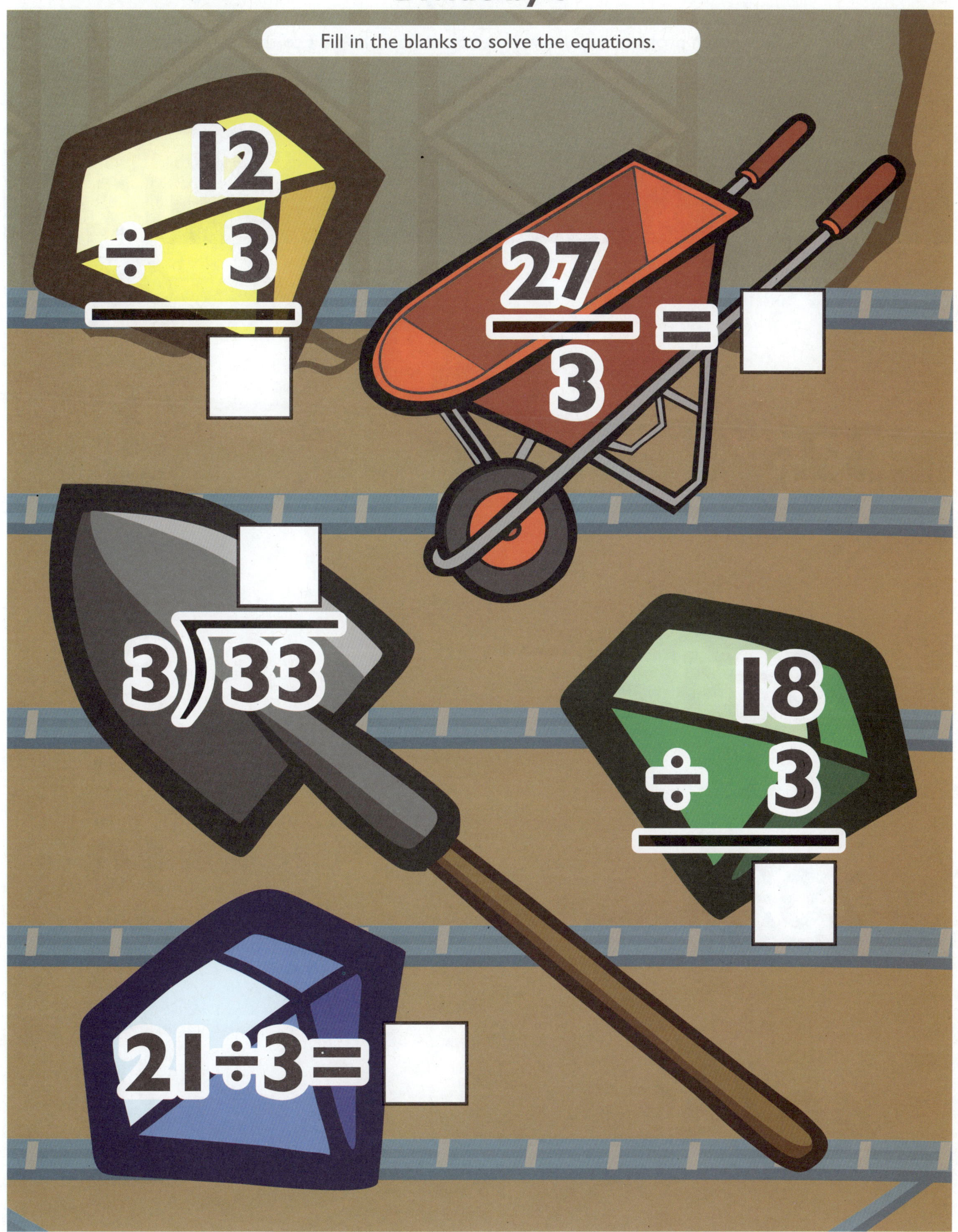

Divide by 4

Solve each problem and draw a line to connect the fruit to the blender.

$4 \div 4$	12
$32 \div 4$	1
$48 \div 4$	5
$20 \div 4$	10
$40 \div 4$	8

Divide by 4

Divide by 5

Solve each problem and draw a line to connect the shells to the bucket.

20÷5

35÷5

10÷5

50÷5

60÷5

Divide by 5

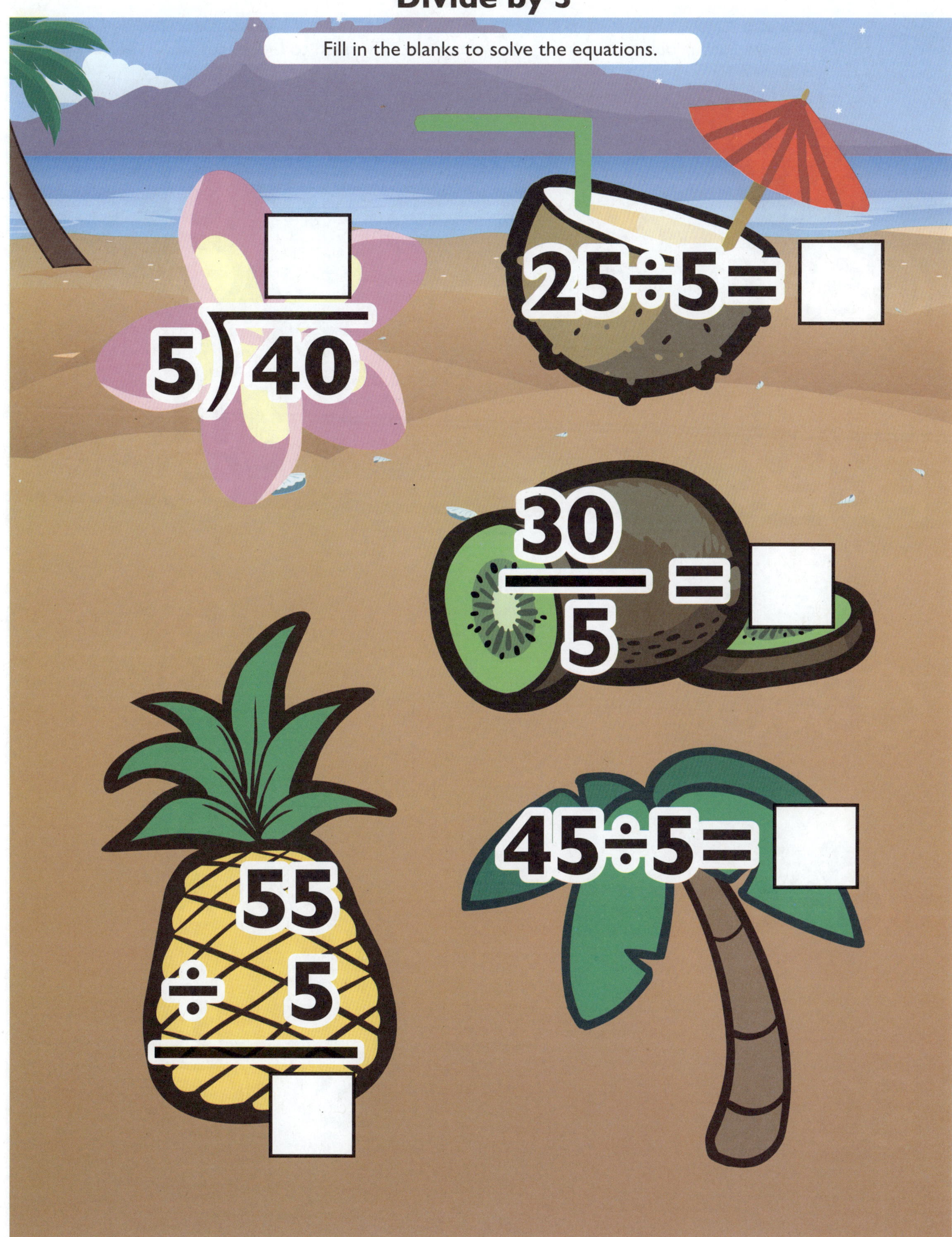

Divide by 6

Solve each problem and draw a line to connect the alien to his spaceship.

Divide by 6

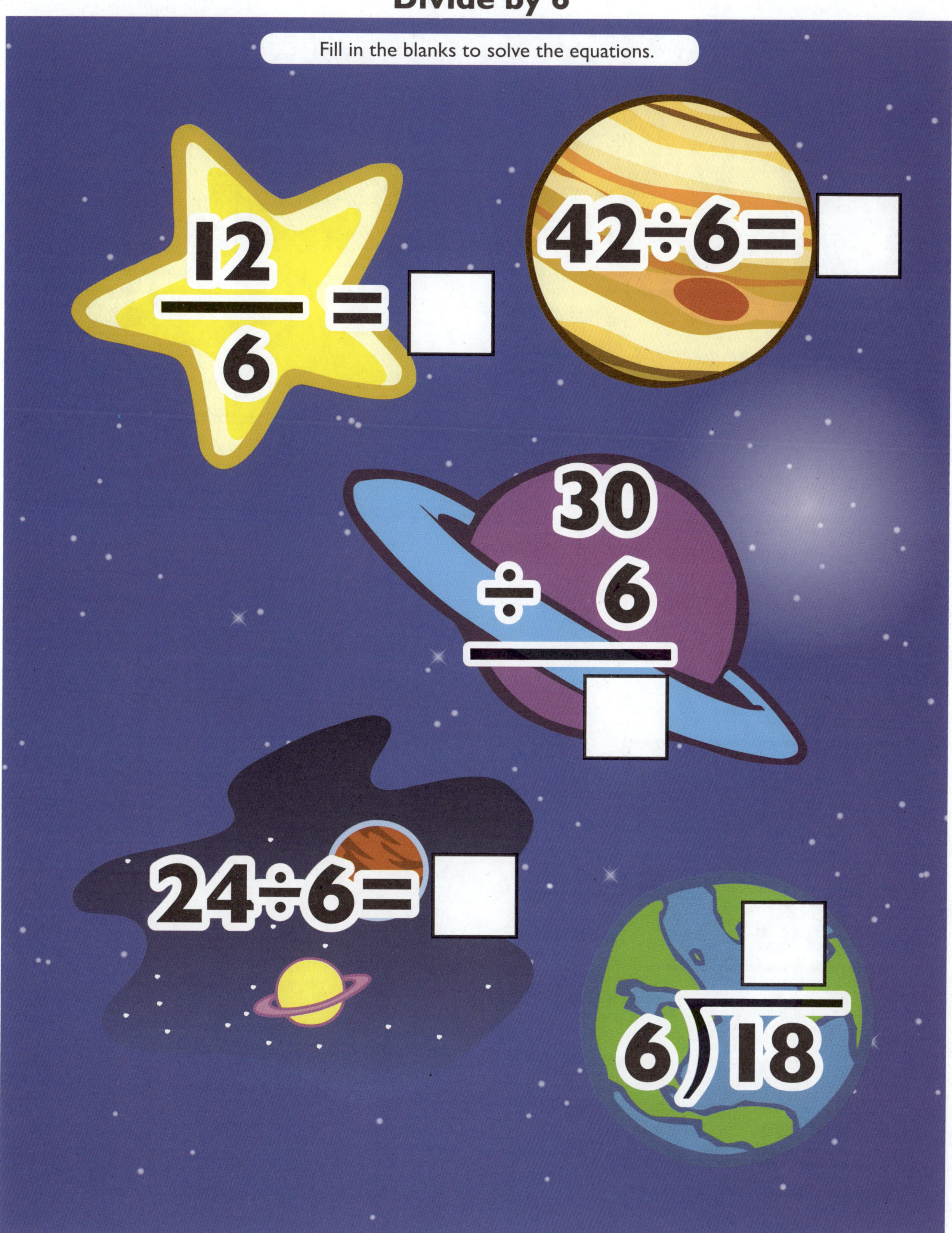

Divide by 7

Solve each problem and draw a line to connect the bow to the violin.
70÷7
42÷7
84÷7
56÷7
49÷7
8
7
10
12
6

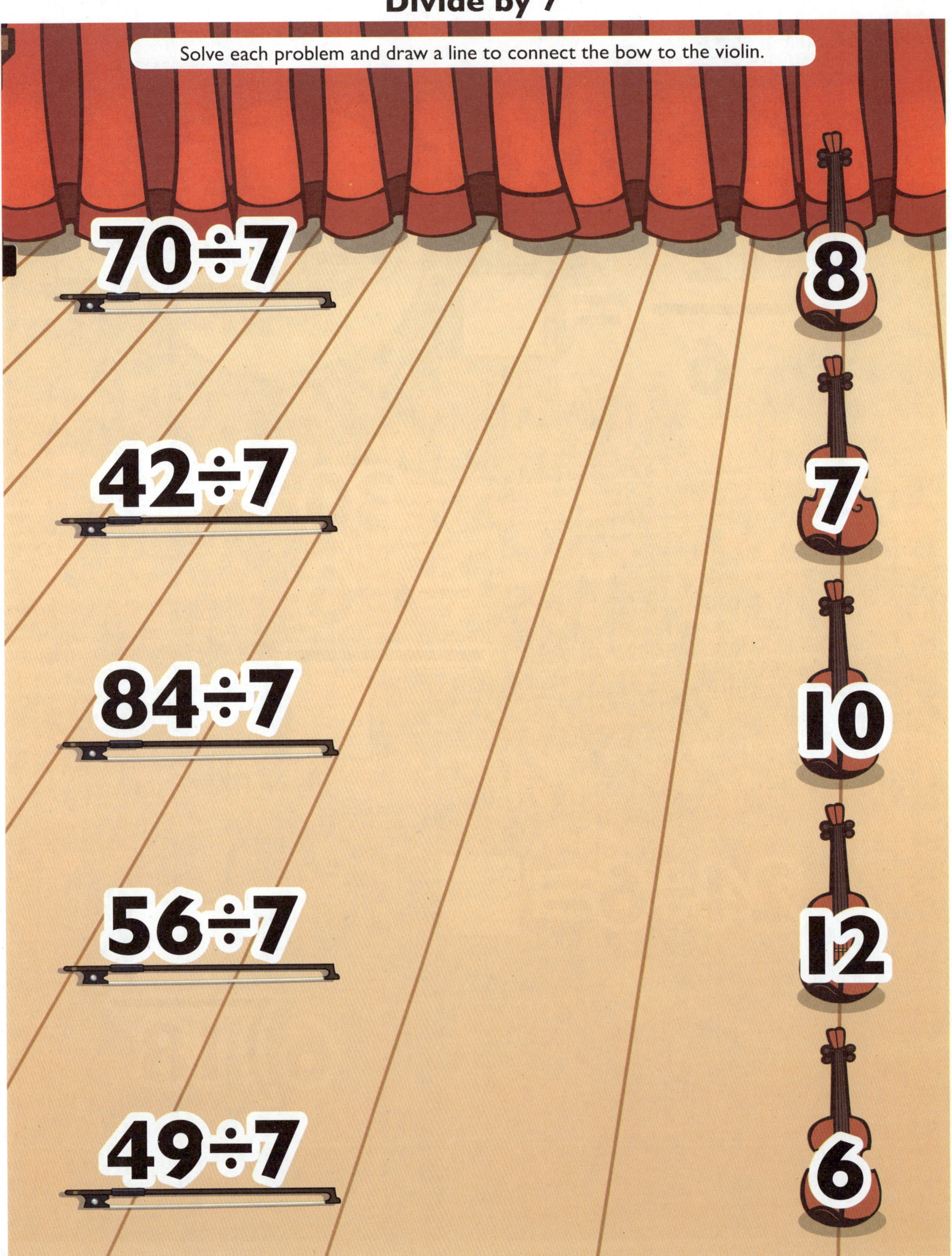

Divide by 7

Divide by 8

Solve each problem and draw a line to connect the ape to the banana.

$64 \div 8$	10
$8 \div 8$	8
$72 \div 8$	1
$80 \div 8$	11
$88 \div 8$	9

Divide by 8

Divide by 9

Solve each problem and draw a line to connect the can to the flowers.

$99 \div 9$	10
$9 \div 9$	12
$90 \div 9$	1
$81 \div 9$	11
$108 \div 9$	9

Divide by 9

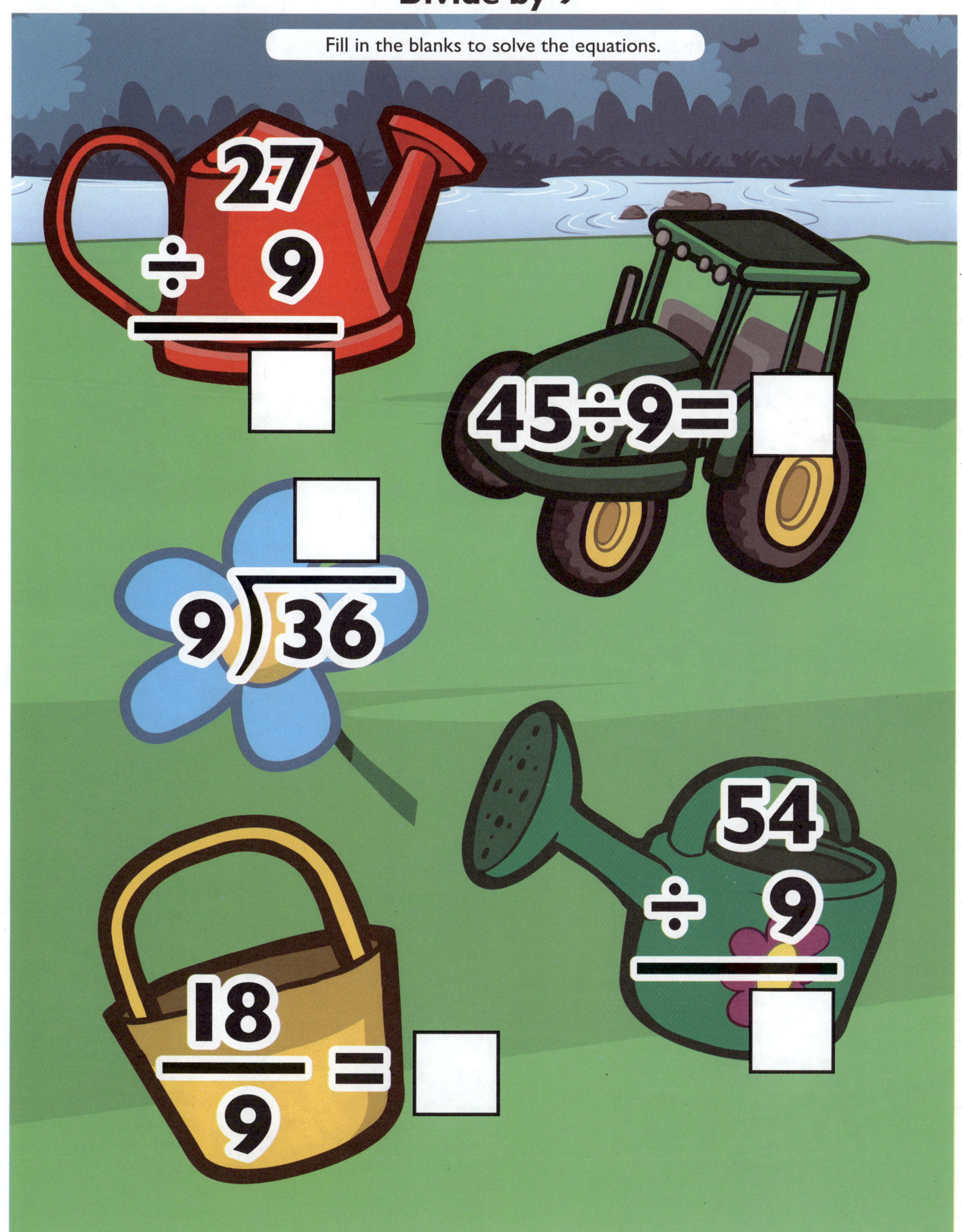

Divide by 10

Solve each problem and draw a line to connect the bird to his nest.

$60 \div 10$	4
$90 \div 10$	7
$40 \div 10$	9
$20 \div 10$	6
$70 \div 10$	2

Divide by 10

Divide by 11

Solve each problem and draw a line to connect the net to the butterfly.

Divide by 11

Fill in the blanks to solve the equations.

Divide by 12

Solve each problem and draw a line to connect the jar to a firefly.

Divide by 12

Fill in the blanks to solve the equations.

Equals 1

Connect the objects that equal 1 to get the rattle to the baby.

Equals 1

Fill in the blanks to solve the equations.

$12 \div 12 = \square$

$2 \div \square = 1$

$9 \div \square = 1$

$8 \div 8 = \square$

$\square \div 5 = 1$

$6 \div \square = 1$

$9 \div \square = 1$

$\square \div 7 = 1$

Equals 2

Equals 2

Equals 3

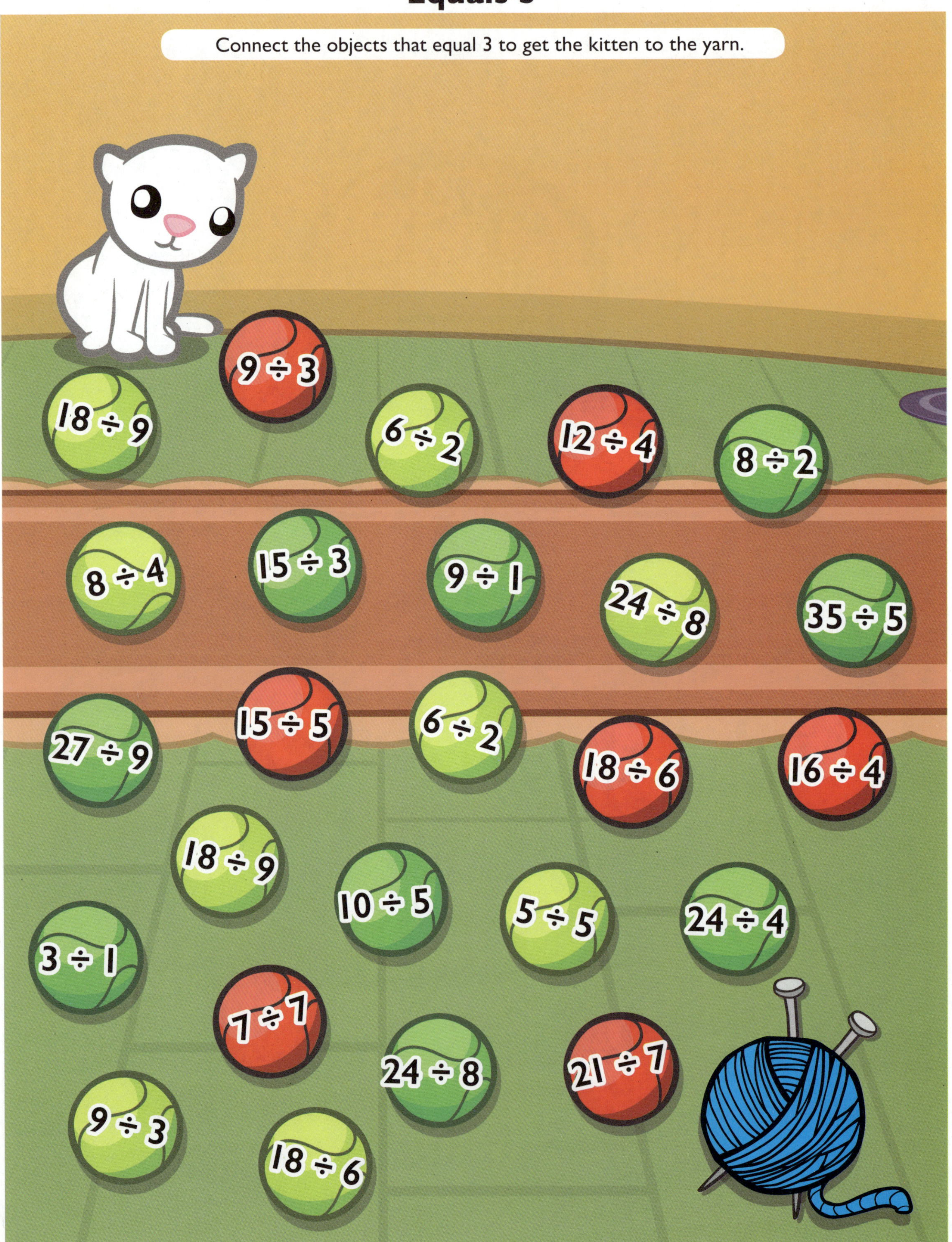

Equals 3

Fill in the blanks to solve the equations.

Equals 4

Equals 4

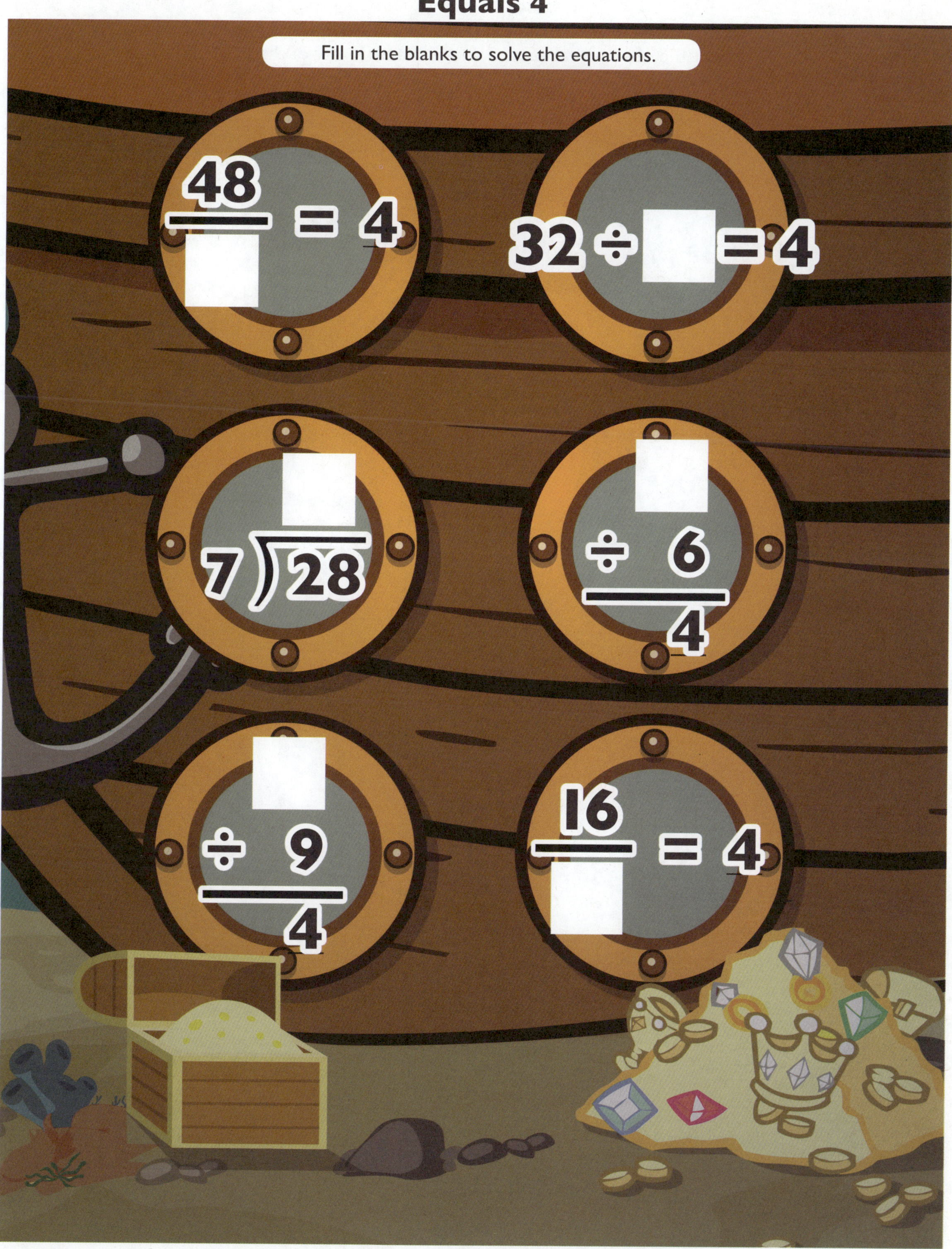

Equals 5

Connect the objects that equal 5 to get the prize out of the machine.

Equals 5

Fill in the blanks to solve the equations.

Equals 6

Equals 6

Fill in the blanks to solve the equations.

Equals 7

Equals 7

Fill in the blanks to solve the equations.

$70 \div 10 = \square$

$\square \div 9 = 7$

$14 \div \square = 7$

$84 \div 12 = \square$

$\square \div 8 = 7$

$77 \div \square = 7$

$21 \div \square = 7$

$28 \div 4 = \square$

$49 \div 7 = \square$

$7 \div \square = 7$

$\square \div 5 = 7$

$\square \div 2 = 7$

Equals 8

Equals 8

Equals 9

Equals 9

Equals 10

Connect the objects that equal 10 to get the mechanics to the car.

Equals 10

Fill in the blanks to solve the equations.

$50 \div 5 = \square$

$20 \div \square = 10$

$\square \div 8 = 10$

$\square \div 3 = 10$

$40 \div \square = 10$

$100 \div 10 = \square$

$\square \div 9 = 10$

$120 \div \square = 10$

$70 \div 7 = \square$

$\square \div 6 = 10$

$80 \div \square = 10$

$30 \div 3 = \square$

Equals 11

Equals 11

Equals 12

Equals 12

Fill in the blanks to solve the equations.

$$\begin{array}{r} \square \\ \div\ 6 \\ \hline 12 \end{array}$$

$$3\overline{)36}\ \text{quotient: } \square$$

$$\begin{array}{r} 24 \\ \div\ \square \\ \hline 12 \end{array}$$

$$\frac{\square}{8} = 12$$

$$\square \div 9 = 12$$

$$\frac{84}{\square} = 12$$

$$\begin{array}{r} \square \\ \div\ 4 \\ \hline 12 \end{array}$$

$$7\overline{)84}\ \text{quotient: } \square$$

$$\begin{array}{r} 48 \\ \div\ \square \\ \hline 12 \end{array}$$

$$\frac{\square}{2} = 12$$

$$\square \div 3 = 12$$

$$\frac{\square}{11} = 12$$

Test Your Skills Divided By 1

$$\begin{array}{r} 3 \\ \div\ 1 \\ \hline \end{array} \quad \begin{array}{r} 8 \\ \div\ 1 \\ \hline \end{array} \quad \begin{array}{r} 12 \\ \div\ 1 \\ \hline \end{array} \quad \begin{array}{r} 9 \\ \div\ 1 \\ \hline \end{array} \quad \begin{array}{r} 1 \\ \div\ 1 \\ \hline \end{array}$$

$$\begin{array}{r} 4 \\ \div\ 1 \\ \hline \end{array} \quad \begin{array}{r} 11 \\ \div\ 1 \\ \hline \end{array} \quad \begin{array}{r} 5 \\ \div\ 1 \\ \hline \end{array} \quad \begin{array}{r} 3 \\ \div\ 1 \\ \hline \end{array} \quad \begin{array}{r} 10 \\ \div\ 1 \\ \hline \end{array} \quad \begin{array}{r} 7 \\ \div\ 1 \\ \hline \end{array}$$

$$\begin{array}{r} 1 \\ \div\ 1 \\ \hline \end{array} \quad \begin{array}{r} 2 \\ \div\ 1 \\ \hline \end{array} \quad \begin{array}{r} 12 \\ \div\ 1 \\ \hline \end{array} \quad \begin{array}{r} 8 \\ \div\ 1 \\ \hline \end{array} \quad \begin{array}{r} 6 \\ \div\ 1 \\ \hline \end{array} \quad \begin{array}{r} 11 \\ \div\ 1 \\ \hline \end{array}$$

$$\begin{array}{r} 3 \\ \div\ 1 \\ \hline \end{array} \quad \begin{array}{r} 10 \\ \div\ 1 \\ \hline \end{array} \quad \begin{array}{r} 4 \\ \div\ 1 \\ \hline \end{array} \quad \begin{array}{r} 9 \\ \div\ 1 \\ \hline \end{array} \quad \begin{array}{r} 5 \\ \div\ 1 \\ \hline \end{array} \quad \begin{array}{r} 2 \\ \div\ 1 \\ \hline \end{array}$$

$12 \div 1 =$ $\qquad$ $9 \div 1 =$ $\qquad$ $1 \div 1 =$

$5 \div 1 =$ $\qquad$ $11 \div 1 =$ $\qquad$ $7 \div 1 =$

$1\overline{)4}$ $\qquad$ $\frac{6}{1} =$ $\qquad$ $1\overline{)10}$ $\qquad$ $\frac{8}{1} =$ $\qquad$ $1\overline{)1}$

$1\overline{)5}$ $\qquad$ $\frac{11}{1} =$ $\qquad$ $1\overline{)3}$ $\qquad$ $\frac{12}{1} =$ $\qquad$ $1\overline{)7}$

Test Your Skills Divided By 2

$24 \div 2$	$2 \div 2$	$10 \div 2$	$22 \div 2$	$4 \div 2$	
$8 \div 2$	$18 \div 2$	$6 \div 2$	$12 \div 2$	$16 \div 2$	$2 \div 2$
$20 \div 2$	$4 \div 2$	$14 \div 2$	$24 \div 2$	$10 \div 2$	$6 \div 2$
$22 \div 2$	$14 \div 2$	$12 \div 2$	$8 \div 2$	$16 \div 2$	$18 \div 2$

$20 \div 2 =$ $6 \div 2 =$ $24 \div 2 =$

$14 \div 2 =$ $18 \div 2 =$ $4 \div 2 =$

$2\overline{)22}$ $\frac{8}{2} =$ $2\overline{)8}$ $\frac{16}{2} =$ $2\overline{)6}$

$2\overline{)16}$ $\frac{12}{2} =$ $2\overline{)20}$ $\frac{24}{2} =$ $2\overline{)14}$

Test Your Skills Divided By 3

$21 \div 3 =$ ___ $12 \div 3 =$ ___ $3 \div 3 =$ ___ $36 \div 3 =$ ___ $18 \div 3 =$ ___

$9 \div 3 =$ ___ $15 \div 3 =$ ___ $24 \div 3 =$ ___ $30 \div 3 =$ ___ $6 \div 3 =$ ___ $27 \div 3 =$ ___

$36 \div 3 =$ ___ $9 \div 3 =$ ___ $18 \div 3 =$ ___ $12 \div 3 =$ ___ $21 \div 3 =$ ___ $3 \div 3 =$ ___

$30 \div 3 =$ ___ $24 \div 3 =$ ___ $27 \div 3 =$ ___ $33 \div 3 =$ ___ $15 \div 3 =$ ___ $6 \div 3 =$ ___

$27 \div 3 =$ $12 \div 3 =$ $9 \div 3 =$

$36 \div 3 =$ $3 \div 3 =$ $15 \div 3 =$

$3\overline{)24}$ $\frac{30}{3} =$ $3\overline{)3}$ $\frac{15}{3} =$ $3\overline{)36}$

$3\overline{)6}$ $\frac{18}{3} =$ $3\overline{)21}$ $\frac{27}{3} =$ $3\overline{)33}$

Test Your Skills Divided By 4

$$\begin{array}{r} 16 \\ \div\ 4 \\ \hline \end{array} \quad \begin{array}{r} 44 \\ \div\ 4 \\ \hline \end{array} \quad \begin{array}{r} 20 \\ \div\ 4 \\ \hline \end{array} \quad \begin{array}{r} 8 \\ \div\ 4 \\ \hline \end{array}$$

$$\begin{array}{r} 24 \\ \div\ 4 \\ \hline \end{array} \quad \begin{array}{r} 4 \\ \div\ 4 \\ \hline \end{array} \quad \begin{array}{r} 32 \\ \div\ 4 \\ \hline \end{array} \quad \begin{array}{r} 48 \\ \div\ 4 \\ \hline \end{array} \quad \begin{array}{r} 28 \\ \div\ 4 \\ \hline \end{array} \quad \begin{array}{r} 12 \\ \div\ 4 \\ \hline \end{array}$$

$$\begin{array}{r} 36 \\ \div\ 4 \\ \hline \end{array} \quad \begin{array}{r} 40 \\ \div\ 4 \\ \hline \end{array} \quad \begin{array}{r} 44 \\ \div\ 4 \\ \hline \end{array} \quad \begin{array}{r} 16 \\ \div\ 4 \\ \hline \end{array} \quad \begin{array}{r} 20 \\ \div\ 4 \\ \hline \end{array} \quad \begin{array}{r} 24 \\ \div\ 4 \\ \hline \end{array}$$

$$\begin{array}{r} 4 \\ \div\ 4 \\ \hline \end{array} \quad \begin{array}{r} 32 \\ \div\ 4 \\ \hline \end{array} \quad \begin{array}{r} 28 \\ \div\ 4 \\ \hline \end{array} \quad \begin{array}{r} 40 \\ \div\ 4 \\ \hline \end{array} \quad \begin{array}{r} 36 \\ \div\ 4 \\ \hline \end{array} \quad \begin{array}{r} 8 \\ \div\ 4 \\ \hline \end{array}$$

$28 \div 4 =$ $\qquad$ $32 \div 4 =$ $\qquad$ $8 \div 4 =$

$44 \div 4 =$ $\qquad$ $16 \div 4 =$ $\qquad$ $20 \div 4 =$

$4\overline{)\,4}$ $\qquad$ $\frac{36}{4} =$ $\qquad$ $4\overline{)\,16}$ $\qquad$ $\frac{32}{4} =$ $\qquad$ $4\overline{)\,44}$

$4\overline{)\,48}$ $\qquad$ $\frac{8}{4} =$ $\qquad$ $4\overline{)\,40}$ $\qquad$ $\frac{4}{4} =$ $\qquad$ $4\overline{)\,20}$

Test Your Skills Divided By 5

$35 \div 5 =$ ___ $15 \div 5 =$ ___ $40 \div 5 =$ ___ $10 \div 5 =$ ___

$45 \div 5 =$ ___ $5 \div 5 =$ ___ $20 \div 5 =$ ___ $55 \div 5 =$ ___ $50 \div 5 =$ ___ $30 \div 5 =$ ___

$60 \div 5 =$ ___ $25 \div 5 =$ ___ $10 \div 5 =$ ___ $35 \div 5 =$ ___ $55 \div 5 =$ ___ $45 \div 5 =$ ___

$5 \div 5 =$ ___ $40 \div 5 =$ ___ $60 \div 5 =$ ___ $15 \div 5 =$ ___ $20 \div 5 =$ ___ $50 \div 5 =$ ___

$40 \div 5 =$ $15 \div 5 =$ $35 \div 5 =$

$5 \div 5 =$ $55 \div 5 =$ $20 \div 5 =$

$5\overline{)55}$ $\frac{50}{5} =$ $5\overline{)35}$ $\frac{45}{5} =$ $5\overline{)40}$

$5\overline{)40}$ $\frac{5}{5} =$ $5\overline{)15}$ $\frac{30}{5} =$ $5\overline{)20}$

Test Your Skills Divided By 6

$$\begin{array}{r} 24 \\ \div \quad 6 \\ \hline \end{array} \quad \begin{array}{r} 72 \\ \div \quad 6 \\ \hline \end{array} \quad \begin{array}{r} 18 \\ \div \quad 6 \\ \hline \end{array} \quad \begin{array}{r} 42 \\ \div \quad 6 \\ \hline \end{array} \quad \begin{array}{r} 54 \\ \div \quad 6 \\ \hline \end{array}$$

$$\begin{array}{r} 66 \\ \div \quad 6 \\ \hline \end{array} \quad \begin{array}{r} 6 \\ \div \quad 6 \\ \hline \end{array} \quad \begin{array}{r} 12 \\ \div \quad 6 \\ \hline \end{array} \quad \begin{array}{r} 48 \\ \div \quad 6 \\ \hline \end{array} \quad \begin{array}{r} 60 \\ \div \quad 6 \\ \hline \end{array} \quad \begin{array}{r} 36 \\ \div \quad 6 \\ \hline \end{array}$$

$$\begin{array}{r} 30 \\ \div \quad 6 \\ \hline \end{array} \quad \begin{array}{r} 72 \\ \div \quad 6 \\ \hline \end{array} \quad \begin{array}{r} 42 \\ \div \quad 6 \\ \hline \end{array} \quad \begin{array}{r} 66 \\ \div \quad 6 \\ \hline \end{array} \quad \begin{array}{r} 54 \\ \div \quad 6 \\ \hline \end{array} \quad \begin{array}{r} 12 \\ \div \quad 6 \\ \hline \end{array}$$

$$\begin{array}{r} 18 \\ \div \quad 6 \\ \hline \end{array} \quad \begin{array}{r} 60 \\ \div \quad 6 \\ \hline \end{array} \quad \begin{array}{r} 36 \\ \div \quad 6 \\ \hline \end{array} \quad \begin{array}{r} 24 \\ \div \quad 6 \\ \hline \end{array} \quad \begin{array}{r} 48 \\ \div \quad 6 \\ \hline \end{array} \quad \begin{array}{r} 6 \\ \div \quad 6 \\ \hline \end{array}$$

$72 \div 6 =$ $\qquad$ $30 \div 6 =$ $\qquad$ $18 \div 6 =$

$48 \div 6 =$ $\qquad$ $66 \div 6 =$ $\qquad$ $54 \div 6 =$

$6\overline{)60}$ $\qquad$ $\frac{12}{6} =$ $\qquad$ $6\overline{)48}$ $\qquad$ $\frac{66}{6} =$ $\qquad$ $6\overline{)54}$

$6\overline{)36}$ $\qquad$ $\frac{60}{6} =$ $\qquad$ $6\overline{)72}$ $\qquad$ $\frac{42}{6} =$ $\qquad$ $6\overline{)30}$

Test Your Skills Divided By 7

$49 \div 7 =$ ___

$21 \div 7 =$ ___

$63 \div 7 =$ ___

$28 \div 7 =$ ___

$14 \div 7 =$ ___

$77 \div 7 =$ ___

$35 \div 7 =$ ___

$84 \div 7 =$ ___

$56 \div 7 =$ ___

$7 \div 7 =$ ___

$42 \div 7 =$ ___

$70 \div 7 =$ ___

$63 \div 7 =$ ___

$35 \div 7 =$ ___

$42 \div 7 =$ ___

$49 \div 7 =$ ___

$28 \div 7 =$ ___

$56 \div 7 =$ ___

$84 \div 7 =$ ___

$7 \div 7 =$ ___

$77 \div 7 =$ ___

$14 \div 7 =$ ___

$21 \div 7 =$ ___

$84 \div 7 =$ $35 \div 7 =$ $77 \div 7 =$

$56 \div 7 =$ $63 \div 7 =$ $21 \div 7 =$

$7\overline{)28}$ $\frac{7}{7} =$ $7\overline{)42}$ $\frac{77}{7} =$ $7\overline{)7}$

$7\overline{)56}$ $\frac{70}{7} =$ $7\overline{)14}$ $\frac{35}{7} =$ $7\overline{)49}$

Test Your Skills Divided By 8

$56 \div 8 =$ ___ $8 \div 8 =$ ___ $80 \div 8 =$ ___ $24 \div 8 =$ ___ $64 \div 8 =$ ___ $88 \div 8 =$ ___

$16 \div 8 =$ ___ $32 \div 8 =$ ___ $72 \div 8 =$ ___ $96 \div 8 =$ ___ $48 \div 8 =$ ___ $40 \div 8 =$ ___

$96 \div 8 =$ ___ $80 \div 8 =$ ___ $32 \div 8 =$ ___ $56 \div 8 =$ ___ $88 \div 8 =$ ___ $48 \div 8 =$ ___

$8 \div 8 =$ ___ $40 \div 8 =$ ___ $16 \div 8 =$ ___ $64 \div 8 =$ ___ $72 \div 8 =$ ___ $24 \div 8 =$ ___

$32 \div 8 =$ $8 \div 8 =$ $24 \div 8 =$

$96 \div 8 =$ $56 \div 8 =$ $64 \div 8 =$

$8\overline{)72}$ $\frac{96}{8} =$ $8\overline{)8}$ $\frac{80}{8} =$ $8\overline{)40}$

$8\overline{)56}$ $\frac{32}{8} =$ $8\overline{)88}$ $\frac{64}{8} =$ $8\overline{)16}$

Test Your Skills Divided By 9

$$\begin{array}{r} 99 \\ \div\ 9 \\ \hline \end{array} \quad \begin{array}{r} 81 \\ \div\ 9 \\ \hline \end{array} \quad \begin{array}{r} 108 \\ \div\ 9 \\ \hline \end{array} \quad \begin{array}{r} 63 \\ \div\ 9 \\ \hline \end{array} \quad \begin{array}{r} 27 \\ \div\ 9 \\ \hline \end{array}$$

$$\begin{array}{r} 90 \\ \div\ 9 \\ \hline \end{array} \quad \begin{array}{r} 36 \\ \div\ 9 \\ \hline \end{array} \quad \begin{array}{r} 9 \\ \div\ 9 \\ \hline \end{array} \quad \begin{array}{r} 72 \\ \div\ 9 \\ \hline \end{array} \quad \begin{array}{r} 45 \\ \div\ 9 \\ \hline \end{array} \quad \begin{array}{r} 18 \\ \div\ 9 \\ \hline \end{array}$$

$$\begin{array}{r} 54 \\ \div\ 9 \\ \hline \end{array} \quad \begin{array}{r} 108 \\ \div\ 9 \\ \hline \end{array} \quad \begin{array}{r} 27 \\ \div\ 9 \\ \hline \end{array} \quad \begin{array}{r} 45 \\ \div\ 9 \\ \hline \end{array} \quad \begin{array}{r} 90 \\ \div\ 9 \\ \hline \end{array} \quad \begin{array}{r} 36 \\ \div\ 9 \\ \hline \end{array}$$

$$\begin{array}{r} 9 \\ \div\ 9 \\ \hline \end{array} \quad \begin{array}{r} 81 \\ \div\ 9 \\ \hline \end{array} \quad \begin{array}{r} 63 \\ \div\ 9 \\ \hline \end{array} \quad \begin{array}{r} 18 \\ \div\ 9 \\ \hline \end{array} \quad \begin{array}{r} 72 \\ \div\ 9 \\ \hline \end{array} \quad \begin{array}{r} 99 \\ \div\ 9 \\ \hline \end{array}$$

$108 \div 9 =$ $\qquad$ $99 \div 9 =$ $\qquad$ $27 \div 9 =$

$81 \div 9 =$ $\qquad$ $36 \div 9 =$ $\qquad$ $90 \div 9 =$

$9\overline{)54}$ $\qquad$ $\frac{63}{9} =$ $\qquad$ $9\overline{)108}$ $\qquad$ $\frac{36}{9} =$ $\qquad$ $9\overline{)99}$

$9\overline{)45}$ $\qquad$ $\frac{81}{9} =$ $\qquad$ $9\overline{)81}$ $\qquad$ $\frac{54}{9} =$ $\qquad$ $9\overline{)9}$

Test Your Skills Divided By 10

$\begin{array}{r} 80 \\ \div\ 10 \\ \hline \end{array}$ $\begin{array}{r} 60 \\ \div\ 10 \\ \hline \end{array}$ $\begin{array}{r} 10 \\ \div\ 10 \\ \hline \end{array}$ $\begin{array}{r} 40 \\ \div\ 10 \\ \hline \end{array}$

$\begin{array}{r} 120 \\ \div\ 10 \\ \hline \end{array}$ $\begin{array}{r} 50 \\ \div\ 10 \\ \hline \end{array}$ $\begin{array}{r} 20 \\ \div\ 10 \\ \hline \end{array}$ $\begin{array}{r} 100 \\ \div\ 10 \\ \hline \end{array}$ $\begin{array}{r} 30 \\ \div\ 10 \\ \hline \end{array}$ $\begin{array}{r} 110 \\ \div\ 10 \\ \hline \end{array}$

$\begin{array}{r} 70 \\ \div\ 10 \\ \hline \end{array}$ $\begin{array}{r} 90 \\ \div\ 10 \\ \hline \end{array}$ $\begin{array}{r} 110 \\ \div\ 10 \\ \hline \end{array}$ $\begin{array}{r} 60 \\ \div\ 10 \\ \hline \end{array}$ $\begin{array}{r} 50 \\ \div\ 10 \\ \hline \end{array}$ $\begin{array}{r} 120 \\ \div\ 10 \\ \hline \end{array}$

$\begin{array}{r} 30 \\ \div\ 10 \\ \hline \end{array}$ $\begin{array}{r} 10 \\ \div\ 10 \\ \hline \end{array}$ $\begin{array}{r} 80 \\ \div\ 10 \\ \hline \end{array}$ $\begin{array}{r} 20 \\ \div\ 10 \\ \hline \end{array}$ $\begin{array}{r} 40 \\ \div\ 10 \\ \hline \end{array}$ $\begin{array}{r} 100 \\ \div\ 10 \\ \hline \end{array}$

$110 \div 10 =$ $50 \div 10 =$ $90 \div 10 =$

$20 \div 10 =$ $100 \div 10 =$ $70 \div 10 =$

$10\overline{)120}$ $\frac{10}{10} =$ $10\overline{)100}$ $\frac{110}{10} =$ $10\overline{)60}$

$10\overline{)20}$ $\frac{70}{10} =$ $10\overline{)40}$ $\frac{30}{10} =$ $10\overline{)10}$

Test Your Skills Divided By 11

$22 \div 11 =$ ___ $55 \div 11 =$ ___ $33 \div 11 =$ ___ $99 \div 11 =$ ___ $121 \div 11 =$ ___

$11 \div 11 =$ ___ $132 \div 11 =$ ___ $44 \div 11 =$ ___ $77 \div 11 =$ ___ $88 \div 11 =$ ___ $110 \div 11 =$ ___

$66 \div 11 =$ ___ $121 \div 11 =$ ___ $22 \div 11 =$ ___ $44 \div 11 =$ ___ $55 \div 11 =$ ___ $88 \div 11 =$ ___

$110 \div 11 =$ ___ $33 \div 11 =$ ___ $11 \div 11 =$ ___ $132 \div 11 =$ ___ $77 \div 11 =$ ___ $99 \div 11 =$ ___

$11 \div 11 =$ $77 \div 11 =$ $110 \div 11 =$

$44 \div 11 =$ $121 \div 11 =$ $33 \div 11 =$

$11\overline{)132}$ $\frac{66}{11} =$ $11\overline{)77}$ $\frac{99}{11} =$ $11\overline{)55}$

$11\overline{)88}$ $\frac{44}{11} =$ $11\overline{)121}$ $\frac{33}{11} =$ $11\overline{)11}$

Test Your Skills Divided By 12

96 ÷ 12	48 ÷ 12	108 ÷ 12	144 ÷ 12

24 ÷ 12	72 ÷ 12	60 ÷ 12	132 ÷ 12	12 ÷ 12	84 ÷ 12
120 ÷ 12	36 ÷ 12	96 ÷ 12	72 ÷ 12	132 ÷ 12	48 ÷ 12
60 ÷ 12	24 ÷ 12	12 ÷ 12	108 ÷ 12	84 ÷ 12	144 ÷ 12

60 ÷ 12 =   144 ÷ 12 =   72 ÷ 12 =

108 ÷ 12 =   48 ÷ 12 =   120 ÷ 12 =

$12\overline{)132}$ $\frac{108}{12} =$ $12\overline{)144}$ $\frac{60}{12} =$ $12\overline{)24}$

$12\overline{)84}$ $\frac{36}{12} =$ $12\overline{)12}$ $\frac{96}{12} =$ $12\overline{)120}$

Mixed Division

$12 \div 4$ $\quad$ $10 \div 5$ $\quad$ $8 \div 8$ $\quad$ $6 \div 2$ $\quad$ $4 \div 1$ $\quad$ $9 \div 3$

$48 \div 12$ $\quad$ $22 \div 2$ $\quad$ $120 \div 10$ $\quad$ $16 \div 8$ $\quad$ $27 \div 3$ $\quad$ $36 \div 6$

$96 \div 12$ $\quad$ $121 \div 11$ $\quad$ $63 \div 9$ $\quad$ $0 \div 12$ $\quad$ $36 \div 4$ $\quad$ $55 \div 11$

$8 \div 2$ $\quad$ $72 \div 6$ $\quad$ $32 \div 8$ $\quad$ $90 \div 9$ $\quad$ $28 \div 7$ $\quad$ $99 \div 11$

$30 \div 5 =$ $\quad$ $56 \div 7 =$ $\quad$ $110 \div 11 =$

$15 \div 3 =$ $\quad$ $110 \div 11 =$ $\quad$ $35 \div 7 =$

$11\overline{)132}$ $\quad$ $\frac{60}{10} =$ $\quad$ $6\overline{)18}$ $\quad$ $\frac{88}{8} =$ $\quad$ $6\overline{)54}$

$2\overline{)10}$ $\quad$ $\frac{15}{3} =$ $\quad$ $12\overline{)132}$ $\quad$ $\frac{54}{6} =$ $\quad$ $2\overline{)22}$

Mixed Division

$100 \div 10$	$48 \div 12$	$99 \div 9$	$144 \div 12$	$2 \div 1$	$84 \div 7$
$49 \div 7$	$6 \div 1$	$44 \div 4$	$18 \div 6$	$25 \div 5$	$45 \div 9$
$120 \div 10$	$36 \div 6$	$72 \div 8$	$77 \div 11$	$110 \div 10$	$64 \div 8$
$24 \div 12$	$54 \div 6$	$12 \div 2$	$121 \div 11$	$60 \div 12$	$144 \div 12$

$30 \div 10 =$ $100 \div 10 =$ $84 \div 12 =$

$21 \div 7 =$ $35 \div 7 =$ $60 \div 5 =$

$11\overline{)132}$ $\frac{120}{10} =$ $9\overline{)108}$ $\frac{45}{5} =$ $9\overline{)81}$

$10\overline{)80}$ $\frac{56}{8} =$ $6\overline{)18}$ $\frac{36}{12} =$ $4\overline{)24}$